WITHDRAWN

A Humane
Economy

A Humane Economy

The Social Framework of the Free Market

By Wilhelm Röpke

Introduction by Dermot Quinn

ISI Books
Intercollegiate Studies Institute
Wilmington, Delaware
1998

Cataloging-in-Publication Data

Röpke, Wilhelm, 1899-1966.

A humane economy: the social framework of the free market/ by Wilhelm Röpke,

— 3rd ed. — Wilmington, DE: Intercollegiate Studies Institute, 1998.

p. cm.

Includes bibliographical references.

ISBN 1-882926-24-2

1. Technology and civilization. 2. Economic policy. I. Title.

HM221 .R6213 1998   98-73066

306/.3—dc21   CIP

Published in the United States by:

Intercollegiate Studies Institute

P.O. Box 4431

Wilmington, DE 19807-0431

www.isi.org

Manufactured in the United States of America

CONTENTS

FOREWORD

Around the turn of the last century, the finial of a church steeple at Gotha was opened. In it was found a document, deposited there in 1784, which read as follows: "Our days are the happiest of the eighteenth century. Emperors, kings, princes descend benevolently from their awe-inspiring height, forsake splendor and pomp, and become their people's father, friend, and confidant. Religion emerges in its divine glory from the tattered clerical gown. Enlightenment makes giant strides. Thousands of our brothers and sisters, who used to live in consecrated idleness, are given back to public life. Religious hatred and intolerance are disappearing, humanity and freedom of thought gain the upper hand. The arts and sciences prosper, and our eyes look deep into nature's workshop. Artisans, like artists, approach perfection, useful knowledge germinates in all estates. This is a faithful picture of our times. Do not look down upon us haughtily if you have attained to greater heights and can see further than we do; mindful of our record, acknowledge how much our courage and strength have raised and supported your position. Do likewise for your successors and be happy." Five years later, the French Revolution broke out; its waves have still not subsided, still throw us hither and thither. Gotha itself, famed for its *Almanach de Gotha* and its sausages, has been engulfed by the most monstrous tyranny of all times.

There could be no greater distance between the honest happiness of the document quoted and the spirit of this book. We may hope, of course, that the German language as written in 1957 would still be intelligible to a burgher of Gotha in 1784. But what, except dumfounded horror, would be his reaction if he were to become acquainted with our world of today—a world shaken by tremendous shocks and menaced by unimaginable disasters, the prey of anxiety, a world adrift and deeply unhappy?

The science of economics had no doubt come to the notice of the erudite in Gotha, thanks to Adam Smith's work, published a few years earlier. But it would seem as incomprehensible as all the rest to our burgher that a representative of that science should be writing a book such as this. Our own contemporaries will comprehend it all the better, in so far as they understand their own situation and the problems of their epoch. To further such understanding is the purpose of this book, as it was the aim of its predecessors. This volume is, however, more than its predecessors were, a book full of apprehension, bitterness, anger and even contempt for the worst features of our age. This is not a sign of the author's growing gloom, but of the progressive deterioration of the crisis in which we live. It is also a book which takes the reader up and down many flights of stairs, through many stories, into many rooms, some light, some dark, into turrets and corners—but that is perhaps the least reproach to be leveled against the author.

What other thoughts I wish to place at the head of this book, I entrust to the French tongue, once more claiming its place as the *lingua franca* of Europe. I could not express these thoughts better than my friend René Gillouin has done in his book *L'homme moderne, bourreau de lui-même* (Paris, 1951): "Ainsi nous sommes tous entraînés dans un courant qui est devenu un torrent, dans un torrent qui est devenu une cataracte, et contre lequel, tant que durera le règne des masses falsifiées, vulgarisées, barbarisées, il serait aussi insensé de lutter que de prétendre remonter le Niagara à la nage. Mais il n'est pas toujours impossible de s'en garer ou de s'en dégager, et alors de se retirer dans ce 'lieu écarté,' dont parle le *Misanthrope* pour y cultiver, dans la solitude ou dans une intimité choisie, loin des propagandes grossières et de leurs mensonges infâmes, lavérité, la pureté, l'authenticité. Que des sécessions de ce genre se multiplient, qu'elles se groupent, qu'elles se fédèrent, elles ne taderont pas à polariser un nombre immense d'esprits droits et de bonnes volontés sincères, qui ont pris le siècle en horreur, mais qui ne savent ni à qui ni à quoi se vouer. Ainsi pourraient se constituer des centres de résistance inviolables, des équipes de fabricants d'*arches* en vue du prochain Déluge, des groupes de reconstructeurs pour le lendemain de la catastophe inéluctable."

—Wilhelm Röpke, Geneva, August, 1957

PREFACE TO THE
ENGLISH-LANGUAGE EDITION

In Dante's time, scholars were, at least in one respect, better off than they are today. They all wrote their books in the same language, namely Latin, and thus did not have to worry about translations. Otherwise, one might surmise that Dante would have reserved to scholars an especially gruesome spot in his *Inferno*: to punish them for their vanity—a failing reputedly not altogether alien to them—they would be made to read translations of their own works into languages with which they were familiar. That this is, as a rule, indeed torture is well known to anyone who has had the experience.

This is the image by which I seek to give adequate expression to the gratitude I owe Mrs. Elizabeth Henderson for the skill and devotion she has brought to the translation of this book, together with her fine feeling for the two languages here to be transposed. What usually is torture for me, she has made a pleasure, and she has lifted me from *Inferno* to *Paradiso*. To be quite honest, it was not an unmitigated pleasure, for she has humbled me by discovering an undue number of errors in the German original. The reader of the English version is the gainer. Indeed, its only essential difference from the German original is the absence of these errors.

I am afraid, however, that even the qualities of this English rendering, while perhaps disposing my critics in the Anglo-Saxon world towards a little more indulgence, will not disarm them. As in the German-speaking world, I expect that my book will meet with four major types of response.

One group of critics will reject the book *en bloc* because it is in flat contradiction with their more or less collectivist and centrist ideas. Another will tell me that in this book, called *A Humane Economy: The Social Framework of the Free Market*, they really appreciate only what is to be found in the world of supply and demand—the world of property—and not what lies beyond. These are the inveterate rationalists, the hard-boiled economists, the prosaic utilitarians, who may all feel that, given proper guidance, I might perhaps have attained to something better. Third, there will be those who, on the contrary, blame me for being a hard-boiled economist myself and who will find something worth praising only in that part of the book which deals with the things beyond supply and demand. These are the pure moralists and romantics, who may perhaps cite me as proof of how a pure soul can be corrupted by political economy. Finally, there may be a fourth group of readers who take a favorable view of the book as a whole and who regard it as one its virtues to have incurred the disapproval of the other three groups.

It would be sheer hypocrisy on my part not confess quite frankly that the last group is my favorite.

—Wilhelm Röpke, Geneva, January, 1960

INTRODUCTION

BY DERMOT QUINN

Economists do not stand high in public esteem, and for good reason. For one thing, they seem to get it wrong as often as they get it right. Offering certainties with the confidence of hard scientists, their predictions dressed in the best mathematical finery, they seem to have a record of success somewhere between that of a fairground madame and a reader of tea-leaves. The public, embarrassed and bemused by this nakedness, rightly prefers the *Farmer's Almanac* to Keynes's *General Theory*. Another problem is that they have power without responsibility. Economists are the unacknowledged legislators of the modern world, the shadowy eminences who determine all our destinies. Yet where are they when their theories explode? No where close to a ballot box. Indeed they not only escape censure but seem to imagine their services to be more indispensable than ever. Like the drunkards in Milton's *Comus*, unaware that wine has made them ugly, they "not once perceive their foul disfigurement/but boast themselves more comely than before." Politicians, not averse to similar boastings of their own, indulge these deceptions. The rest of us have to clear up the glasses. Yet these complaints only touch the surface of the problem. Mostly the public resents not so much the economist's showy obscurantism—graphs, formulas and the like—as the behaviorism that lies behind it. Like ghosts in someone else's machine, we are reduced to the measurable, the predictable, the banal. Not much humanity remains when all our strivings, our work and play, are known in advance and forecast with a slide-rule. Thus a paradox. The economist likes to speak of rational choice, but having conferred upon us the dignity of reason he renders it useless by describing *homo oeconomicus* as nothing more than an amalgam of rational choices. Ortega y Gasset once wrote of the expulsion of man from art. In *A Humane Economy* Wilhelm Röpke penned a brilliant denunciation of the expulsion of man from economics. The proper measure of mankind is man, said Alexander Pope. Man is also the proper measure of economics. "It is the precept of ethical and humane behavior, no less than of political wisdom," Röpke wrote, "to adapt economic policy to man, not man to economic policy." That, in sum, is the central insight of the book.

The point is simple enough: Economics is a dismal science precisely because it claims to be a science. It turns human effort into a quadratic equation. Yet it was not always this way. Adam Smith, first of the moderns in economic thinking, would have deplored the mathematical aridities of his neo-liberal disciples, who, like Oscar Wilde's cynic, seem to know the price of everything and the value of nothing. Smith was a moral philosopher before he was an economist. For him, the wealth of nations was not to be measured simply in money but in all of those social excellences that promote human flourishing: neighborliness, community, family, self-reliance, provision for the future. Economic activity removed from moral understanding was inconceivable to him. After all, buying and selling presumes some measure of trust between individuals, a disposition towards honorable mutual advantage. He would have found it strange that someone might write a book called *A Humane Economy* as if there could be any other kind. If Smith's moral theory is in some ways flawed—there is more psychology than philosophy in it—at least he recognized that economics does not stand alone but forms part of a broader understanding of the human person. Some of his followers disagree, generally with baneful results. The declension of economics into calculation and calibration remains the besetting weakness of the subject. Art should not ape science; morality should not mimic mechanics.

Human dignity is, then, the central concern of economics. Yet to read *A Humane Economy* is to be made aware, with a terrible insistency, that precisely this dignity is threatened by the social and economic arrangements of our day. Consider the phenomenon which Röpke termed *vermassung* or "enmassment," the sheer crowdedness of contemporary life:

As we increasingly become merely passively activated mass particles or social molecules, all poetry and dignity, and with them the very spice of life and its human content, go out of life. Even the dramatic episodes of existence—birth, sickness and death—take place in collectivized institutions.... People live in mass quarters, superimposed upon each other vertically and extending horizontally as far as the eye can see; they work in mass factories or offices in hierarchical subordination; they spend their Sundays and vacations in masses, flood the universities, lecture halls and laboratories in masses, read books and newspapers printed in millions and of a level that usually corresponds to these mass sales, are assailed at every turn by the same billboards, submit, with millions of others, to the same movie, radio and television programs.... Only the churches are empty, almost a refuge of solitude....

This grim sociology of vastness, with its emphasis on collective over individual experience, largeness over smallness, self-indulgence over self-reliance, tends to fragment the human personality if it does not obliterate it entirely. The herd must be soothed, rendered passive, by magazines, television, gadgets and assorted electronic trinketry. The muzak of the malls ought to be Bach: "Sheep may safely graze." That is modernity's anthem.

Such jeremiads may seem excessive. Progress has its problems, it might be argued, but these are unavoidable in a crowded planet. Anyway is this not a price worth paying for universal material prosperity? Yet distaste for "the noise and stench of mechanized mass living" is not—as some might think—elitism or cultivated contempt for the herd. Nor is it pre-industrial romanticism, although, as Röpke pointed out, the cult of the standard of living amounts to a profound spiritual disorder. The point is not the banality of mass culture— although that is even more evident than when Röpke wrote—as its degraded moral sensibility. In the age of the masses, proletarianized man has chosen that others should make choices for him. Decisions are forever shifted upwards, from the individual, the family, and the group to corporations, federations, the state itself. There is a kind of moral infantilism in such dependency. But there is also, as with infants, a curious petulance. Children can be willful, as if aware from time to time of their own weakness. Mass man, for all his shop-bought opinions and second-hand attitudes, congratulates himself on his own individualism. And indeed, Röpke argued, there is a dangerously individualistic strain in modernity. Detached from tradition, community, and family, from the moorings of natural law, modern man is left to his own devices. Rootless, he embraces any roots. Isolated, he seeks any society. Disintegrated, he craves any creed that might make him whole. "Loneliness, separateness, and isolation," suggested Röpke, "are becoming the destiny of the masses." It is a situation pathological to the point of collective lunacy.

The lineaments of a social vision begin to emerge. Röpke stood squarely in the Burkean tradition that celebrated "little platoons"—family, church, community, civic association, rootedness in local things—as the best bulwark against the "bloated colossus" of the state. "The freedom of society resides in its pluralism," he argued, not in the tyranny of uniformity and centralization. This localism was not the smothering smallness that some critics see in conservatism, the elevation of the merely old and familiar into political principle. On the contrary it was a recognition that human societies must have human identity. Röpke did not propose miniaturist or naive social anthropomorphism so much as simple common sense. What was the alternative? The state as some Leviathan of rationalist planning that destroyed its subjects, then itself. That kind of totalitarianism, he suggested, is a defining characteristic of the age of the masses: the government assumes duties not its own and discharges them poorly, confiscates money not its own and disburses it prodigally, claims rights not its own and exercises them prejudicially. The collectivist experiment may begin with high ideals, but it ends by infantilizing the citizenry, creating a pocket-money state and a bureaucracy of nannies to run it.

Röpke was Burkean in another way. In the thought of both men there is an elegiac quality, a longing for lost virtue, a desire for recovery. The "unbought grace of life" is gone, lamented Burke: decency, honor, heroism are despised. Röpke also spoke of those unbought graces—nature, privacy, beauty, dignity, and repose which are the gifts that lie beyond prosy utilitarianism. This sorrowful tone causes trouble for conservatives. It smacks of nostalgia and cultural despair, self-pity and snobbery compounded into a flimsy philosophy of regret. *A Humane Economy* is certainly suffused with a sense of loss. Consider its autobiographical evocation of life before the age of the masses. "A village and small-town childhood, with its confident ease, its plenty, and its now unimaginable freedom and almost cloudless optimism" gave way, Röpke wrote, to war, revolution, inflation, depression, and unemployment. The dichotomy seems overdone. Yet it is accurate enough. Röpke's world *did* disintegrate in 1914: so did that of millions of others.

All the same, nostalgia is the weakest foundation for conservatism, no matter how historically plausible it may be. But nostalgia is precisely what Röpke did *not* offer. He did not wish to set the clock back: rather, as he said, he wished to set it right. To do so he offered specific remedies for specific problems, all of them technically sophisticated, economically well thought out. But what kind of society did he seek to restore? It was one in which

wealth would be widely dispersed: people's lives would have solid foundations; genuine communities, from the family upward, would form a background of moral support for the individual; there would be counterweights to competition and the mechanical operation of prices; people would have roots and not be adrift in life without an anchor; there would be a broad belt of an independent middle class, a healthy balance between town and country, industry and agriculture.

This has a familiar ring. If *A Humane Economy* owes a debt to Burke, it also draws on Chesterton, Belloc, and the English distributist tradition. Compare, for instance, Belloc's distaste for monopoly capitalism and its proletarianization of the masses with Röpke's distaste for the "concentration" of wealth that allowed government, administration and wealth to be controlled by the powerful few at the expense of the powerless many. Monopoly, he said, was the worst form of commercialism, "economically...questionable and morally...reprehensible." He might almost have been quoting from *The Servile State*. There are other similarities. *A Humane Economy* condemned the politics of "anonymous, unchallengeable and inscrutable" decisions, the inevitable concomitant of "concentration." *The Servile State* decried the "tendency to govern by clique" which "could not possibly arise in a genuinely democratic society." The list could be extended. Both men recognized the insolence and fraudulence of plutocracy, whether it speaks with an English or German accent.

If "concentration" is the problem, dispersal, it would seem, ought to be the answer. Both Röpke and the distributists held that private property widely diffused was the principal foundation of political freedom, the surest defense against the self-aggrandizing state. When property ceases to be a natural and primary right and is held instead at the whim of government, then "the end of free society is in sight." Why? Because "property" enshrines notions of privacy, independence, self-reliance, freedom, and dignity which are principles that tyranny cannot abide. Ownership at once presupposes individual rights and promotes them. Röpke saw it as the cornerstone of the humane economy; Chesterton saw it as the outline of sanity; Belloc saw its absence as characteristic of a servile state. Its primary purpose and true merit lay beyond the visible and measurable. Property broadly understood represented a "particular philosophy of life,...a particular social and moral universe."

This plea for property was a plea against proletarianization. Only through ownership—of house, or land, or skill, or savings—could workingmen free themselves from dependency. (Welfare is the most obvious dependency but wage-slavery is another.) But it was also, in a more positive sense, a plea for embourgeoisement. Property implies the existence of a society in which "certain fundamentals are respected and color the whole network of social relationships." Those foundations are:

> individual effort and responsibility, absolute norms and values, independence based on ownership, prudence and daring, calculating and saving, responsibility for planning one's own life, proper coherence with the community, family feeling, a sense of

tradition and the succession of generations combined with an open-minded view of the present and the future, proper tension between individual and community, firm moral discipline, respect for the value of money, the courage to grapple on one's own with life and its uncertainties, a sense of the natural order of things, and a firm scale of values.

Property, in short, is social propriety. Without it the particular moral universe of bourgeois values simply ceases to exist.

Yet if property is a trust it also requires trust. It enshrines a particular relationship between the citizenry and the state. To be sure, the state is more than an insurance arrangement between owners. Hobbesian or Lockean Man is not the best face that humanity can collectively muster. Nonetheless a key function of the state is fiduciary. It must protect property. When it fails to do so it embarks upon its own dissolution, even if—the irony is obvious—it conceals that fact by ever greater acts of enlargement and bureaucratic centralization. Citizens may reasonably expect that governments should be fiscally responsible and financially sound; that property should be freely heritable; that its use may be enjoyed primarily by themselves and their beneficiaries; that it should be protected from predatory taxation and predatory inflation; that it should not be redistributed, with prodigies of moral self-regard, to fund welfarism that is itself morally corrosive. Likewise easy credit should be avoided: it is improvidence in another form. Saving should be encouraged. Regulation should be reduced. Progressive taxation should be pared back. The private should be preferred over the public. Monopolies should be dismantled. Banks should defend money, not dispense it at every turn. Small businesses should be promoted. Inflation should be fought without quarter.

These may seem like poujadist imperatives: shopkeeper economics, nothing more. That is hardly alarming. They have formed the basis of sound political economy for centuries and are not false for being unfashionable. To be sure, Röpke wrote against the tenor of his times, confident of later vindication. In obvious ways, that vindication has come. Keynes, his nemesis, has been exposed and deposed; the state is a little less bloated than before. Yet to read *A Humane Economy* in a spirit of smugness is manifestly to miss the point. The struggle against collectivism is never entirely won: it must be waged again and again in every generation. But it must be fought with proper understanding of the nature of the struggle. Free markets are preferable to tyranny not because they enrich us but because they moralize us. They connect us to authentic human communities, allowing us to be self-reliant yet also honorably dependent on the efforts of others. And precisely for that reason, to make a *cult* of the market is to detach it from its own moral imperatives. Markets do not generate moral norms: they presume them. Moreover, they offer the freedom of self-discipline, not unanchored greed. Besides, for all their excellence, markets are not everything. The vital things, Röpke realized, are those that lie beyond supply and demand, beyond the world of property, beyond the calculator's reach. Such is the point and paradox of the book. A humane economy is only, in the end, a shadowy reflection of the divine one.

To make a government requires no great prudence. Settle the seat of power; teach obedience: and the work is done. To give freedom is still more easy. It is not necessary to guide; it only requires to let go the rein. But to form a free government; *that is, to temper together these opposite elements of liberty and restraint in one consistent work, requires much thought, deep reflection, a sagacious, powerful, and combining mind.*

—EDMUND BURKE, *Reflections on the Revolution in France,* 1790

Men are qualified for civil liberty in exact proportion to their disposition to put moral chains upon their own appetites; in proportion as their love of justice is above their rapacity; in proportion as their soundness and sobriety of understanding is above their vanity and presumption; in proportion as they are more disposed to listen to the counsels of the wise and good, in preference to the flattery of knaves. Society cannot exist unless a controlling power upon will and appetite be placed somewhere, and the less of it there is within, the more there must be without. It is ordained in the eternal constitution of things, that men of intemperate minds cannot be free. Their passions forge their fetters.

—EDMUND BURKE, *A Letter from Mr. Burke to a Member of the National Assembly in Answer to Some Objections to His Book on French Affairs,* 1791

CHAPTER I

Reappraisal After Fifteen Years

Personal

About fifteen years ago I undertook the task of organizing into
something like a coherent whole my ideas and opinions on *The
Social Crisis of Our Time*, as my book was called. That first outline
was subsequently filled in by two more rather detailed works, *Civitas Humana* and *Internationale Ordnung—heute*, which also appeared nearly as long ago, while even the papers later brought together in *Mass und Mitte* are now some ten years old. Much has
happened since then, much has been thought and written, and the
political, economic, and spiritual development of society since the
collapse of National Socialist totalitarianism has been somewhat
stormy. I feel now that it is incumbent upon me to take up once
more my original subject, and to do so in a manner which will
bring out what is permanently valid in the various topical and fragmentary contributions which I have tried to make in the meantime
to the discussion of the old questions as well as of some new ones.

What has happened in those fifteen years, and just where do we
stand today? What is to be said now in the context of the problems
discussed in *The Social Crisis of Our Time?* These are the questions which first come to mind. They are questions to which one
individual seeks an answer—the author of those earlier works and
of this one—and they are questions to which it is not possible to
give any but a subjective reply, however much it may be based on

1

arguments made as cogent as possible and on the widest possible experience. It is, therefore, both honest and expedient for the author to begin with himself and to try to define his own position in regard to social and economic affairs. If this achieves nothing else, it may at least set an example.

Those who, like myself, were born a few weeks before the close of the last century can regard themselves as coevals of the twentieth century, although they cannot hope to see its end. Anyone who has, as I have, the somewhat doubtful privilege of having been born a national of one of the great powers and, moreover, of one of the most turbulent powers of this great, tragic continent and who has shared its varied fate throughout the phases of his life may add, like millions of others, that his experience spans a wider range than is normally given to man. A village and small-town childhood which, with its confident ease, its plenty, and its now unimaginable freedom and almost cloudless optimism, was still set in the great century of liberalism that ended in 1914 was followed by a world war, a revolution, and crushing inflation; then came a period of deceptive calm, followed by the Great Depression, with its millions of unemployed; then a new upheaval and an eruption of evil when the very foundations of middle-class society seemed to give way and the pathetic stream of people driven from house and home ushered in the new age of the barbarians; and finally, as the inescapable end to this appalling horror, another and more terrible world war. Now, before we have even taken the full measure of the political, economic, social, and spiritual shocks which this war engendered, the world is menaced by the sole surviving, the Communist, variety of totalitarianism and by the apocalyptic prospects of unleashed atomic energy.

What has been the impact of this experience and of its interpretation on a man like myself? Perhaps the one thing I know most definitely is something negative: I can hardly describe myself as a socialist in any meaningful or commonly accepted sense. It took me a long time to become quite clear on this point, but today it

2

seems to me that this statement, properly understood, is the most clear-cut, firm, and definite part of my beliefs. But this is where the problem begins. Where does a man of my kind take his stand if he is to attack socialism because he believes it to be wrong?

Is the standpoint of liberalism the right one to deliver his attack? In a certain sense, yes, if liberalism is understood as faith in a particular "social technique," that is, in a particular economic order. If it is liberal to entrust economic order, not to planning, coercion, and penalties, but to the spontaneous and free co-operation of people through the market, price, and competition, and at the same time to regard property as the pillar of this free order, then I speak as a liberal when I reject socialism. The technique of socialism—that is, economic planning, nationalization, the erosion of property, and the cradle-to-the-grave welfare state—has done great harm in our times; on the other hand, we have the irrefutable testimony of the last fifteen years, particularly in Germany, that the opposite—the liberal—technique of the market economy opens the way to well-being, freedom, the rule of law, the distribution of power, and international co-operation. These are the facts, and they demand the adoption of a firm position *against* the socialist and *for* the liberal kind of economic order.

The history of the last fifteen years, which is that of the failure of the socialist technique all along the line and of the triumph of the market economy, is indeed such as to lend great force to this faith. But, if we think it through, it is much more than simple faith in a social technique inspired by the laws of economics. I have rallied to it not merely because, as an economist, I flatter myself that I have some grasp of the working of prices, interest, costs, and exchange rates. The true reason lies deeper, in those levels where each man's social philosophy is rooted. And here I am not at all sure that I do not belong to the conservative rather than the liberal camp, in so far as I dissociate myself from certain principles of social philosophy which, over long stretches of the history of thought, rested on common foundations with liberalism and social-

3

ism, or at least accompanied them. I have in mind such "isms" as utilitarianism, progressivism, secularism, rationalism, optimism, and what Eric Voegelin aptly calls "immanentism" or "social gnosticism."[1]

In the last resort, the distinction between socialists and non-socialists is one which divides men who hold basically different views of life and its true meaning and of the nature of man and society. Cardinal Manning's statement that "all human differences are ultimately religious ones" goes to the core of the matter. The view we take of man's nature and position in the universe ultimately determines whether we choose man himself or else "society," the "group," or the "community" as our standard of reference for social values. Our decision on this point becomes the watershed of our political thinking, even though we may not always be clearly aware of this and may take some time to realize it. This remains true in spite of the fact that in most cases people's political thinking is by no means in line with their most profound religious and philosophical convictions because intricate economic or other questions mask the conflict. People may be led by Christian and humane convictions to declare themselves in sympathy with socialism and may actually believe that this is the best safeguard of man's spiritual personality against the encroachments of power, but they fail to see that this means favoring a social and economic order which threatens to destroy their ideal of man and human freedom. There remains the hope that one may be able to make them aware of their error and persuade them by means of irrefutable, or at least reasonable, arguments that their choice in the field of economic and social order may have consequences which are diametrically opposed to their own philosophy.

As far as I myself am concerned, what I reject in socialism is a philosophy which, any "liberal" phraseology it may use notwithstanding, places too little emphasis on man, his nature, and his personality and which, at least in its enthusiasm for anything that may be described as organization, concentration, management, and ad-

4

ministrative machinery, makes light of the danger that all this may lead to the sacrifice of freedom in the plain and tragic sense exemplified by the totalitarian state. My picture of man is fashioned by the spiritual heritage of classical and Christian tradition. I see in man the likeness of God; I am profoundly convinced that it is an appalling sin to reduce man to a means (even in the name of high-sounding phrases) and that each man's soul is something unique, irreplaceable, priceless, in comparison with which all other things are as naught. I am attached to a humanism which is rooted in these convictions and which regards man as the child and image of God, but not as God himself, to be idolized as he is by the *hubris* of a false and atheist humanism. These, I believe, are the reasons why I so greatly distrust all forms of collectivism.

It is for the same reasons that I champion an economic order ruled by free prices and markets—and also because weighty arguments and compelling evidence show clearly that in our age of highly developed industrial economy, this is the only economic order compatible with human freedom, with a state and society which safeguard freedom, and with the rule of law. For these are the fundamental conditions without which a life possessing meaning and dignity is impossible for men of our religious and philosophical convictions and traditions. We would uphold this economic order even if it imposed upon nations some material sacrifice while socialism held out the certain promise of enhanced well-being. How fortunate for us it is, then, that precisely the opposite is true, as experience must surely have made obvious by now, even to the most stubborn.

Thus we announce the theme which is the red thread running through this book: the vital things are those beyond supply and demand and the world of property. It is they which give meaning, dignity, and inner richness to life, those purposes and values which belong to the realm of ethics in the widest sense. There is a profound ethical reason why an economy governed by free prices, free markets, and free competition implies health and plenty, while the

5

socialist economy means sickness, disorder, and lower productivity. The liberal economic system releases and utilizes the extraordinary forces inherent in individual self-assertion, whereas the socialist system suppresses them and wears itself out in opposing them. We have—as we shall have occasion later to show in detail—every reason to distrust the moralizing attitudes of those who condemn the free economy because they regard the individual's attempts to assert and advance himself by productive effort as ethically questionable and prefer an economic system which summons the power of the state against them. We are entitled to set least store by such moralizing attitudes when they are preached by intellectuals who have the open or secret ambition to occupy positions of command in such an economic system but who are not critical enough of themselves to suspect their own, ethically none too edifying, *libido dominandi*. They want to use the horsewhip to drive the carriage of virtue through impracticable terrain, and they fail to see that it is downright immorality to lead people into temptation by an economic order which forces them to act against their natural instinct of self-assertion and against the commands of reason. A government which, in peacetime, relies on exchange control, price control, and invidious confiscatory taxation has little, if any, more moral justification on its side than the individual who defends himself against this sort of compulsion by circumventing, or even breaking, the law. It is the precept of ethical and humane behavior, no less than of political wisdom, to adapt economic policy to man, not man to economic policy.

In these considerations lies the essential justification of ownership, profit, and competition. But—and we shall come back to this later—they are justifiable only within certain limits, and in remembering this we return to the realm beyond supply and demand. In other words, the market economy is not everything. It must find its place within a higher order of things which is not ruled by supply and demand, free prices, and competition.

Now nothing is more detrimental to a sound general order appropriate to human nature than two things: mass and concentration. Individual responsibility and independence in proper balance with the community, neighborly spirit, and true civic sense—all of these presuppose that the communities in which we live do not exceed the human scale. They are possible only on the small or medium scale, in an environment of which one can take the measure, in conditions which do not completely destroy or stifle the primary forms of human existence such as survive in our villages and small- or medium-sized towns.

We all know to what a pass we have come in this respect. There is no doubt that what, even fifteen years ago, sounded to many people like fruitless nostalgia for a lost paradise is today a lone voice competing, without hope, against a hurricane. In all fields, mass and concentration are the mark of modern society; they smother the area of individual responsibility, life, and thought and give the strongest impulse to collective thought and feeling. The small circles—from the family on up—with their human warmth and natural solidarity, are giving way before mass and concentration, before the amorphous conglomeration of people in huge cities and industrial centers, before rootlessness and mass organizations, before the anonymous bureaucracy of giant concerns and, eventually, of government itself, which holds this crumbling society together through the coercive machinery of the welfare state, the police, and the tax screw. This is what was ailing modern society even before the Second World War, and since then the illness has become more acute and quite unmistakable. It is a desperate disease calling for the desperate cure of decentralization and deproletarianization. People need to be taken out of the mass and given roots again.

Here, too, lies one of the basic reasons for the crisis of modern democracy, which has gradually degenerated into a centralized mass democracy of Jacobin complexion and which stands more urgently than ever in need of those counterweights of which I spoke in my book *Civitas Humana*. Thus we are led to a political view whose conservative ingredients are plainly recognizable in our predilection

7

for natural law, tradition, *corps intermédiaires*, federalism, and other defenses against the flood of modern mass democracy. We should harbor no illusions about the fateful road which leads from the Jacobinism of the French Revolution to modern totalitarianism.

Nor should we deceive ourselves about all the forces of spiritual and moral decay that are at work everywhere in the name of "modern living" or about the claims which this magic formula is naïvely supposed to justify. I have named some of these forces in my book *Mass und Mitte*, where I had harsh words to say about such movements as progressivism, *sinistrismo*, rationalism, and intellectualism. We should know where all this will end unless we stop it in time. It is no use seeking salvation in institutions, programs, and projects. We shall save ourselves only if more and more of us have the unfashionable courage to take counsel with our own souls and, in the midst of all this modern hustle and bustle, to bethink ourselves of the firm, enduring, and proved truths of life.

This brings me to the very center of my convictions, which, I hope, I share with many others. I have always been reluctant to talk about it because I am not one to air my religious views in public, but let me say it here quite plainly: the ultimate source of our civilization's disease is the spiritual and religious crisis which has overtaken all of us and which each must master for himself. Above all, man is *Homo religiosus*, and yet we have, for the past century, made the desperate attempt to get along without God, and in the place of God we have set up the cult of man, his profane or even ungodly science and art, his technical achievements, and his State. We may be certain that some day the whole world will come to see, in a blinding flash, what is now clear to only a few, namely, that this desperate attempt has created a situation in which man can have no spiritual and moral life, and this means that he cannot live at all for any length of time, in spite of television and speedways and holiday trips and comfortable apartments. We seem to have proved the existence of God in yet another way: by the practical consequences of His assumed non-existence.

8

Surely, no one who is at all honest with himself can fail to be struck by the shocking dechristianization and secularization of our culture. We may try to seek comfort in remembering that this is not the first time that Christianity has ceased to be a living force. As long ago as the eighteenth century it looked, mainly in France but to some extent also in England, as if the Christian tradition and Christian faith had irretrievably lost their hold. It is important and interesting to remember this, and it reminds us that today's lack of religion has its historical origin in eighteenth-century deism or open atheism. But this is cold comfort. However much the content of the Christian tradition was diluted at that time, most people, with all their skepticism and secularism, still believed in a divine order and in a life whose meaning extended beyond this world. But in our times, the situation is radically different. Although, in contrast to the situation in the eighteenth century, a minority takes its faith more seriously than ever and gathers round the equally steadfast churches, the dominating and prevailing opinion is completely atheistic. And since men obviously cannot live in a religious vacuum, they cling to surrogate religions of all kinds, to political passions, ideologies, and pipe dreams—unless, of course, they prefer to drug themselves with the sheer mechanics of producing and consuming, with sport and betting, with sexuality, with rowdiness and crime and the thousand other things which fill our daily newspapers.

We may find some comfort in the reflection that in all this we are still reaping that which destructive spirits of the past have sown. This is still the direction, we may say, in which the same old spirit—or non-spirit—blows, and why should the wind not change, and quite soon, too?

It may do so. We certainly do not wish to exclude the possibility. However, and here we return again to our main theme, we would merely be deluding ourselves if we drew such a sharp dividing line between the realm of the spirit and the conditions of man's existence. We must not shirk the serious question of whether the forms of our modern urban and industrial society are not in themselves a breeding ground for the godlessness and animalism of our times. "Certainly," a contemporary German writer says, "there is a connection between a people's degree of civilization and the degree of its religious faith. In nature, God is revealed to us and we feel His breath; in the cities we are surrounded by man-made things. And the more the man-made things pile up and crowd out nature and natural things, the more we lose the ability to hear the voice of God. Absorbed in the contemplation of a garden, Luther once remarked that man is not capable of making a single rose. It is an obvious remark, and yet a significant one, which expresses a sense of the difference between God's handiwork and man's. In the open country, with the starry skies above our head and the brown fertile earth under our feet, we breathe in God's strength at every step. The countryman, whose work is dictated by the changing seasons and is dependent on the elements, feels himself to be a creature of the Almighty like the corn in the field and the star following its predestined orbit. The ancient invocations of the Psalms come to his lips spontaneously as if but newly spoken. He knows in his heart how far beyond reach God is and at the same time how intimately close, how unfathomable His will and His mercy. It may be that we could precisely calculate the relation between the decline of true faith and the rise of urban civilization cut off from nature if we knew more about such mental processes as faith."[2]

There can be no doubt that the progress of our civilization has meant a steady expansion of the man-made area. On the spiritual plane, it has meant the concurrent spread of the belief that anything which can be extolled as modern is true progress and that there is no limit to what man can "make." If we include in these limitless possibilities man himself as a spiritual and moral being, as well as human society and economy, we step right into Communism. No one would deny that it becomes ever harder to hear God's voice in this man-made, artificial world of ours, on either side of the bank cashier's counter, in the factory, amid the columns of motor traffic and the concrete canyons of our cities, not to mention the underground cities of the full-blown atomic age, like a vision of hell worthy of Brueghel or Bosch, which a German scientist not long ago cheerfully prognosticated for us as the next stage of life on this planet. Worse still, we are getting into the habit of screening off all the vital things, like birth and sickness and death, and we even deprive death of its dignity and moving solemnity by rushing our funeral processions in motorcars through the busy streets to distant cemeteries, those human refuse heaps that are prudently removed, indeed hidden, from residential areas. Who can still have the courage to speak of the flight from the land, for instance, as a desirable by-product of an ever improving pattern of production?

Old and New Vistas

It is quite terrifying to see how people, and not least their spokesmen in public, remain insensitive and criminally optimistic in the face of the social and cultural crisis of our times. If anything, the crisis is getting worse rather than better, and the danger of exaggerating it seems incomparably smaller than that of minimizing it with deceptive, soothing words. If there is a feeling that this crisis is something historically unprecedented and monstrous, that all the standards and precepts of experience are failing us, and that we can no longer count on any constants in human nature or on such unshakable convictions as have hitherto given meaning to our civilization and cultural universe, then this feeling of having the ground cut from under one's feet, of being uprooted and at sea without an anchor, deserves respect and not derision.

Is it really so wide of the mark to think our epoch unique, unprecedented, and unparalleled? Is it not made plain to us at every turn that there are things in our time which never were before, indeed things which determine our whole life?

In fact, it is precisely the decisive elements of our life today that

11

have never existed before in all the millennia of human history, back to its remotest sources, which we are now learning to trace. Never before was there such an increase in the world's population as has taken place during the last two centuries and still continues. Never before have the world's peoples so consciously realized their oneness and, indeed, taken it for granted. Never before has one particular form of civilization become so universal and spanned the whole world as does our Western civilization. Never before has there been anything like the technical triumphs originating in the Western world.[3] And the consequences of modern technology, like modern technology itself, have no precedent in the whole course of known history; they are consequences which affect the life and thought of all those millions whose pattern of existence is set by teeming cities and huge industrial concerns and proletarianism. There finally arises the question of whether all this revolutionary novelty in the external conditions of life is not also accompanied by a radical transformation of the prevailing human type, a new mutation of *Homo sapiens* that is as novel and original as New York and General Motors, radio and nuclear fission, mass living and a world-wide engineering culture.[4]

It is a poor species of human being which this grim vision conjures up before our eyes: "fragmentary and disintegrated" man, the end product of growing mechanization, specialization, and functionalization, which decompose the unity of human personality and dissolve it in the mass, an aborted form of *Homo sapiens* created by a largely technical civilization, a race of spiritual and moral pygmies lending itself willingly—indeed gladly, because that way lies redemption—to use as raw material for the modern collectivist and totalitarian mass state. At the same time, this new type of man is spiritually homeless and morally shipwrecked. His capacity for true religious faith and for cherishing cultural traditions having been worn away by too much cerebral and psychoanalytic introspection, he looks for surrogates in the fanatically intolerant political and social ideologies of our time (or "social religions," as

Alfred Weber aptly calls them). At the top of the list are socialism, Communism, and nationalism.

No one has the right to make light of these dismal forebodings. Nevertheless, we may well ask ourselves whether this kind of pessimism is not so extreme that it becomes part of our cultural crisis and as such must be overcome if we are to find a way out. I am the last to deny the importance of external conditions of living as fashioned by technical progress, organization, and social institutions, but the ultimately decisive things for man lie in the deeper spiritual and moral levels. To assume that external conditions alone determine man's spiritual and moral make-up, and thereby his whole personality, is to concede one of the major aspects of the cultural crisis, namely, the dissolution of our traditional Christian and humanistic conception of man in a sort of historical relativism which defines man in terms of evolutionary stages, morphological types, and cultural cycles. The essential symptom of our cultural crisis is precisely that we are losing the inner certainty which the Christian and humanistic belief in the unity of civilization and man gave us. To overcome the crisis means to regain this certainty. We can save ourselves only if man finds the way back to himself and to the firm shore of his own nature, assured value judgments, and binding faith.[5] At the same time, let there be no doubt about it: man must, of course, also master the immense problems for which we have to thank the unexampled upheaval in our external forms of living.

Therefore, if we must guard against the optimism which does not even suspect the quicksands surrounding us, we must equally guard against the pessimism which sinks in them and, indeed, becomes a new danger and may prove itself true by its own effects on the cultural crisis. We have to beware of an historicism which dissolves everything into change and evolution, as well as of a kind of relativistic sociology which cannot but weaken our position still further. It would be self-contradiction to fail to associate myself with the warning against the extreme danger threatening man and society

13

today, and, to be sure, the causes of danger include some weighty ones previously unknown in history. But this is no reason to allow ourselves to be cast down by these apocalyptic visions. This is not the first time that danger has beset mankind. Nothing compels us to believe that we cannot overcome it, provided only that we hold fast to one ultimate belief: faith in man, faith in man's essentially unalterable nature, and faith in the absolute values from which human dignity derives.

These reflections are highly topical. We can test them at once against an event which, in the midst of all our troubles, we are entitled to enter on the credit side of the balance sheet of our hopes and fears. I have in mind the decline of Communism as a spiritual and moral world power. The colossal monument of totalitarianism left standing after the destruction of National Socialism is crumbling. True, the external military power of Communism is more redoubtable than ever, now that the Russians have gained a start in rocket techniques, and its internal police power remains as revolting as ever. Yet if we correctly interpret the development of the Communist empire since Stalin's death and assess at their true value recent events, not only in satellite countries but in Russia itself, then we can hardly doubt that the Communist doctrine, faith, and appeal have entered a phase of decline, at least in Europe, though less so in the underdeveloped countries of Asia and Africa. It is, of course, too early to speak of collapse, but for the first time in some thirty years we may at least hope, with legitimate confidence, that the end will come. It would be fatal, however, to misinterpret that hope and to allow it to deflect us from the utmost firmness, vigilance, and resolution in the face of Communism as a dictatorship bent on world conquest or to succumb to the seductions of "co-existence" which build upon our gullibility, cowardice, and confusion. On the contrary, what this now justified hope should do is to give us back our long-lost courage and the confidence that we can win that battle for our own existence which Communism has forced upon the free world.

14

This is the vantage point from which we should look, above all, at the chain of dramatic events in the Communist empire which shook the world not long ago. The defiance of the Poles and the open anti-Communist revolt of the heroic Hungarians were stupendous events never to be forgotten. Communism here suffered a moral defeat without our having lifted a finger and, indeed, with many of us in the Western world not even properly knowing what it was all about. It was a defeat whose implications cannot be overestimated, however much naked force seems to prevail once more. The painted stage-sets of Communism have collapsed; it stands unmasked and in such a manner that it cannot hope to recover, however pessimistic a view we may take of the West's forgetfulness and apathy. If things have come to such a pass that a world movement which poses grandiloquently as the liberator of the masses has to treat workers, peasants, and students as its worst enemies, then this is the beginning of the end. It is a resounding defeat in the "Third World War," in which we have long been engaged on terms chosen by Moscow and Peking on the basis of momentary expediency. What is more, the scene of the defeat was the Communists' favorite front, that of internal softening up. The smoke screen of "co-existence," behind which our downfall was being prepared, has blown away; the popular-front intrigues and the whisperings of contact men have become contemptible; gone, above all, is any faith in the Communist "remaking of man," which was to have started with the young. The fund of confidence which Moscow had begun to accumulate in innocent minds with the dexterity of a marriage swindler is suddenly devalued. All we have to do is to make sure that this success is not frittered away by folly, forgetfulness, cowardice, and insensibility. These are indeed already busy doing their worst, and they are aided and abetted by a certain kind of intellectual cleverness which is the very opposite of wisdom.

There remains the essential fact that what we witnessed was the revolt of whole peoples against the violation of the human soul, the darkest and most dangerous aspect of Communism. That man can

15

be "remade" has always been one of the principal tenets of Communism and one worthy of its extreme anti-humane doctrine. The fainthearted or perplexed in the non-Communist world despondently shared this belief for a long time without realizing that they were thereby betraying the Christian and humane conception of the nature of man. Those of my readers who are familiar with my *Mass und Mitte* will recall that however pessimistic I may have been at that time, I always did the best I could to oppose and fight this insolent and degrading doctrine as a mechanistic and atheist denial of human dignity. The events which I have mentioned here are a triumph for our confidence in the essential human soul. Nothing could have been more unequivocal and convincing. The Communist "remaking" was, naturally enough, to have started with the young, but it was the young themselves, who had grown up with the trickery of Communism and had been fed at its trough, who hurled themselves against the Russian tanks most bravely, desperately, and implacably. Even in those parts of the Communist empire where, unlike Hungary, no revolt has yet broken out, it is the rising generation which is the stronghold of spiritual resistance to Communism.

And so we have gained this supreme certainty: whatever disasters Communism may still inflict upon the world, not least because of our own weakness, it will go the way of all godless effrontery. It will tremble before the rebellion of those who fight for freedom and human dignity and who spew the venom of this doctrine out of their mouths. This certainty rests on the conviction, confirmed by experience, that there is a limit to even the worst pessimism with which we may be inclined to view the present crisis of civilization and society. This pessimism is bounded by the primary constants of human nature. These we may trust, but such trust must include a profession of faith in the inviolable core of human nature, as well as the resolution to defend this faith against all corrosive doctrines.

However, having stated these limits and so erected a last bulwark against the philosophy of self-annihilation and despair, we are at

16

once compelled to pronounce another warning against underestimating the extraordinary gravity of the situation. True, there is an end to all things and so to Communism also, and it is comforting to know that even the Russians cannot always have everything their own way. This is a cheering thought and we will give it its due. True, also, the star of Communism as a substitute for religion is declining, and therefore the prospect of our being engulfed forever in the night of totalitarianism has lost much of its terror. But it would be fatal to draw too sanguine conclusions. The theory of the internal decay of Communism can all too easily be abused as a sop to our conscience and in extenuation and excuse of cowardice, fear, and confusion. And all this at a time when that part of the world which calls itself free is threatened as much as ever, and perhaps more, by perils requiring our relentless vigilance and effort, even if the danger of Communism be reduced to one of military and political power which, though still formidable, can be mastered.

It is not the specter of totalitarianism which raises its Gorgon's head in our midst. What we have to fear is a development which, like inflation, is a creeping process and, as will be shown, is indeed closely connected with the creeping inflation of our times. Security and personal comforts are rated more highly than freedom, law, and personality. That which still goes under the name of freedom is, as often as not, license, arbitrariness, laxity, and unlimited demands. Few people today attach to the word "freedom" any clear meaning which might put them on guard against its demagogic abuse. The individual means less and less, mass and collectivity more and more—and so the net of servitude which hems in personal development becomes ever denser, more closely meshed, and inescapable. The center of gravity of decisions keeps shifting upwards: from the individual, the family, and the small, compact group up to anonymous institutions. The power of the state grows uncontrollably; yet, since powerful forces are at the same time eroding its structure and weakening the sense of community, there is less and less assurance that administration and legislation unswervingly

17

serve the whole nation and its long-term interests. Demagogy and pressure groups turn politics into the art of finding the way of least resistance and immediate expediency or into a device for channeling other people's money to one's own group.

Government, legislation, and politics of this kind are bound to forfeit public esteem and to lose their moral authority. They are alarmingly exposed to contempt, lawlessness, and lack of public spirit and to corruption in all forms and degrees. Those who serve government, legislation, or politics experience, in their turn, a corresponding loss of esteem, and this impairs their capacity for resistance to the forces undermining the state. The vicious circle closes. It is all the more vicious since, at the same time, the sheer power of the state, the area of legislation, and the influence of politics have grown enormously and are still expanding, even where governments have assumed office with the explicit promise of lightening this burden. Law and the principles of law, which should be as a rock, become uncertain. For Locke, the individual had an inalienable right to life, liberty, and property; but today, even in the "free world," the last-named pillar is badly dilapidated, and few realize that its fall would pull down the pillar of liberty as well and that the remaining pillar, the right to the inviolability of life, could not then stand alone.

If property is degraded into precarious *de facto* possession depending on the whim of government (shocking cases of this kind have recently occurred in Locke's own country and are, indeed, the writing on the wall) or on the voter's favor; if property becomes a hostage in the hands of those who own less or nothing; if property, together with its inseparable concomitant, the law of inheritance, ceases to be one of the natural and primary rights which need no other justification than that of law itself—then the end of free society is in sight. When governments no longer regard their own nationals' property as sacrosanct, then it is hardly surprising that their rough treatment of foreign property exceeds all the bounds of law and propriety. Few things have so hampered Western govern-

18

ments in dealing with the breaches of contract of Asian or African rulers as the irrefutable fact that nationalization in the West has set an example for the Mossadeghs and Nassers and Sukarnos. Once the relations between nations are ruled no longer by respect for property but by arbitrariness and contempt of law, then the last foundations of international order are threatened and the ubiquitous social crisis extends to the international community. The economic consequences of such a crisis of confidence inevitably fall most heavily upon those underdeveloped countries which are responsible for it, for they are cutting themselves off from Western capital sources, without which economic development can proceed only at great sacrifice.

The crisis of the state and the weakening of common purposes combine with the disregard of property rights to create a situation which is causing more and more concern to the nations of the free world: the diminishing of the value of money by inflation. On the one hand it is not at all certain that governments, in the position where mass society, with its pathological symptoms, has put them, are still able or even willing to stop the rot of their currencies. It is high time that we face the question of whether this problem has not grown out of hand and marks the point where the inner weakness of overblown government is first and most dangerously displayed. On the other hand, no great perspicacity is needed to recognize the close kinship between lack of respect for property and indifference to the value of money. Erosion of property and erosion of money go together; in both cases, that which is solid, stable, firmly held, assured and meant to last is replaced by that which is brittle, precarious, fleeting, uncertain, and meant for the day.

Both kinds of erosion, too, promote each other. Not only are they the result of the same forces, but it can also be shown—and we shall do so later—that the erosion of property and the disintegration of an order of society resting on property create, in many ways, a most favorable climate for the erosion of money by inflation. Conversely, it stands to reason that the erosion of money by

19

inflation strengthens the forces which undermine the position of property because the masses come to lose both the desire and the ability to accumulate individual property and concentrate instead on the flow of income guaranteed by the welfare state and full employment. It is true that inflation induces people to put their money into real values, but on the other hand, it exposes property to all the shocks and social tensions which are peculiar to an inflationary climate and widens the gap between the victims of inflation and those who know how to defend themselves. So, once more, we move in a fateful spiral from which no easy escape is now possible, least of all by the reckless optimism of those who refuse to face the facts and problems of the crisis.

Market Economy and Collectivism

With these considerations we have reached the more particular field of economics, which provides the occasion of conducting our reappraisal in a manner befitting the economist. The first question is this: To what extent and with what degree of lasting success has that economic order which is appropriate to a free society, namely, the market economy, been able to hold its own against the collectivist economic order, which is incompatible with free society in the long run?

At first sight it may seem as if the adherent of the market economy had good reason to feel both satisfied and hopeful when he considers the conflict between these two principles of economic order during the last fifteen years. He may feel all the more entitled to do so if he recalls the straits in which the cause of the market economy found itself when the Second World War and its outcome seemed to clinch the triumph of collectivism throughout the world. Founded upon the corresponding collectivist ideologies, controlled and planned economies, with their paraphernalia of forms, fixed prices, rationing, injunctions and permits, police checks, and penalties, seemed to be holding the field all along the line. When

20

the Second World War was drawing to its close, few had the courage to give the market economy a good character, let alone a future. This small band was headed by a handful of men who had asked themselves long ago what were the fundamental difficulties on which the collectivist economy, with its central administration and compulsion, was bound to come to grief, and what incomparable advantages the market economy had on its side. Long before the market economy was again taken for granted and had become the source of prosperity, as they expected, these men in various countries had set to work to popularize the idea of economic order as a system of regulatory principles and incentives in the economy and to explain that in the last resort the choice lies between only two systems: the collectivist system, resting on planning and commands—as Walter Eucken says, the "economy of central administration"—and the opposite system of the market economy.

In our forgetful era it may be useful to recall how poor the prospects for the market economy appeared at that time and how hopeless the efforts of its advocates. What was the situation at the end of the Second World War? Throughout a whole century one of the principal reasons for the advance of socialism had been the myth of its historical necessity, with which Marx, above all, had equipped the movement. This myth was well adapted to the mental inertia of the man in the street, and its propaganda value was bound to increase when the day of fulfillment actually seemed to be at hand. It is hard to withstand the appeal of an idea which is not only the winner-designate in the timetable of history, known only to the initiates, but which actually seems to have won through already. This is exactly what the situation was then.

Nearly everywhere in the world the purposes of planning, nationalization, and full employment had given rise to a mixture of expansionist monetary policy and official controls that paralyzed the price mechanism. The Leftist course of economic policy, with its varieties in different countries, owed its ascendancy in part to Keynes's oft-misunderstood ideas and in part to the heritage of war

21

and war economy. The triumph was also furthered by the fiction that the Allied victory over the Fascist countries was tantamount to the victory of an anti-Fascist (that is, predominantly socialist and progressive) front over the bloc of powers mistaken for ultra-conservatives, reactionaries, and monopoly capitalists. The blindness with which collectivist totalitarian Russia was accepted as a member of this anti-Fascist front was matched by stubborn refusal to accept any proof that German National Socialism had, at least in a formal sense, paralleled Soviet Russia as a textbook example of socialism in full bloom and had, on the spiritual side, more than one ancestor in common with "democratic" socialism. Those who, like F. A. Hayek and the author of this book, were so deplorably tactless as to explode this myth know from experience what it means to challenge a popular misconception.[6]

Only in the light of all this can we appreciate the full significance of the fact that gradually a number of European countries began to form a center of opposition and had the temerity to disregard the timetable of history. In 1945, Switzerland stood alone as a kind of museum piece of liberalism which could be dismissed with an indulgent smile. The first jolt came when, in 1946, Belgium followed in the tracks of Switzerland and set her economy on an even keel by stopping inflation and reintroducing a free-market system. She was soon so successful that her balance-of-payments equilibrium disqualified her from direct Marshall Plan aid, which was tailored to the needs of socialist countries. At the same time, Sweden, which had started out from a situation comparable to Switzerland's, effectively demonstrated that determined Leftist policies, inspired by socialist theoreticians, enable even a rich country, and one spared by war, to soften the hardest currency almost overnight. But, the obtuse might have argued, did not Belgium possess the riches of the Congo, which would explain the miracle without destroying the socialist and inflationary creed? The answer was not long in coming. By adopting the now famous policy of Luigi Einaudi, then Governor of the Bank of Italy and later President of the Republic,

22

who put a professor's knowledge of economics into practice, Italy, in 1947, rallied to the nucleus of liberal and anti-inflationary countries and managed to extricate herself from the morass of inflation and economic controls. It was a striking success and most probably saved Italy from the victory of Communism; however, its demonstration value was somewhat overshadowed by the host of problems peculiar to Italy.

The really decisive victory in the critical European economic situation was won by Germany in the summer of 1948. Again it was a professor who switched from theory to practice. Ludwig Erhard and his group, stepping into a situation of so-called repressed inflation which was really nothing less than the stark and complete bankruptcy of inflationary collectivism, countered with a resolute return to the market economy and monetary discipline. What is more, Erhard was unsporting enough to succeed beyond all expectations.[7] This was the beginning of an impressive chapter in economic history when in the span of a few years we witnessed a nation's precipitous fall and its rebirth and the almost total collapse and subsequent swift recovery of its economy. The world was treated to a unique and instructive example of the paralysis and anarchy which can afflict an economy when utterly mistaken economic policies destroy the foundations of economic order and of how quickly and thoroughly it can recover from its fall and start on a steep, upward climb if only economic policy recognizes its error and reverses its course.

Germany lay prostrate, ravaged by war, impoverished by ten years of repressed inflation, mutilated, demoralized by defeat in an unjustified war and by the exposure of a hateful tyranny, and teeming with refugees: a country without hope which the passengers of international express trains traversed hastily, embarrassed by the children scrounging for leftovers of food on the embankments. And of all countries, it was precisely this one which had the courage to swim against the tide of collectivist and inflationary policies in Europe and set up its own—and contrary—program of free markets and monetary discipline, much to the dismay of the young economists of the Occupation Powers, who had been reared on the doctrines of Marx and Keynes and their disciples. Not only was the success overwhelming, but it also happened to coincide with the patent failure of socialism in Great Britain (which had replaced the hopelessly discredited Soviet Union as the promised land of the socialists). It began to look as if the country that had lost the war were better off than the winners.[8]

This was an outrage because it meant the end of the socialist myth. What made it still harder to swallow was that defeated Germany should be the one to set this example of prosperity through freedom, for there were few who grasped that this was a not unworthy manner of making good some of the evil that this same country had brought upon the world by means of its previous, opposite example of inflationary collectivism, which had marked the beginning of National Socialism and had been lapped up all too eagerly by others. But the success of the new economic course was proscribed by every single chapter of the fashionable Leftist economic doctrine. This success simply could not be tolerated, and thus false theories combined with wishful thinking to produce those repeated prophecies of doom which accompanied German economic policy from one triumph to the next. When the false prophets, along with all their various disproved predictions, finally fell into ridicule, they took refuge in the tactics of either remaining as silent as possible on the success of the German market economy or of obscuring it with all kinds of statistical juggling and gross misrepresentation of facts. They also liked to dwell upon such problems as still required solution, exaggerating their importance and unjustly blaming the market economy for them. The annual reports of the United Nations Economic Commission for Europe in Geneva are a treasure-trove in this respect.

The dire warnings began with the assertion that truncated West Germany was not economically viable. This theme was repeated in a minor key in all sorts of variations until the symphony of disaster had to be broken off abruptly when it became obvious that Germany had become one of the world's leading industrial and trading nations, the major economic power on the Continent, and the possessor of one of the hardest and most sought-after currencies. Other arguments were then brought up. Some said that it was all a flash in the pan, others that currency reform and Marshall Plan aid were the good fairies rather than the market economy. Germany was a black sheep of sinister economic reaction and crippling deflation, the worst problem child of Europe besides Belgium and Italy, or so the annual reports of the European Economic Commission would have one believe. Still others said that even if Germany were quite obviously prospering, it had nothing to do with the market economy; it was only because the Germans were such a hard-working, frugal, and thrifty people. There is really no point now in continuing this list of embarrassed evasions and absurdities because they have long been answered by the facts. The lesson which Germany (and later Austria, in equally difficult conditions, and most recently France) taught the world with its example of market economy and monetary discipline gradually emerged from the sphere of demagogic disparagement and ideological party strife and became one of the most important reasons why the market economy has put collectivism on the defensive everywhere this side of the Iron Curtain—and, indeed, outside Europe, too, as the impressive example of Peru has shown.

If we look back today upon the economic development of the major countries of the West since the Second World War, the story is one of grave economic debility at first and subsequent recovery. True, the recovery is far from being complete, nor can we be certain that it will be lasting, but all the same, it has produced impressive improvements. The disease was caused by the economic experiments of dabblers and an unholy alliance of inflation and collectivism. The recovery, to the extent that it has taken place, is, we repeat, attributable to a very simple prescription: the re-establishment of a workable and stable currency system and the liberation

25

of the economy from the fetters of planning, which had hampered or entirely paralyzed the regulatory and incentive effects of free prices and free competition. In some countries, especially Germany, the recovery was as unparalleled as the preceding sickness, and it is this recovery which opened the way to the spectacular revival of international economic relations, especially within Europe.

Under the impact of these experiences, the dispute over the principles of economic policy in a free society has now become much less acute. Nationalization and planning, the catch phrases of the immediate postwar period, have lost their appeal, and even in the socialist camp the response is now weak. Such enthusiastic reception as they still find—together with nationalism, which is closely connected with them—seems to be concentrated in the so-called underdeveloped countries, with their Nehrus and Sukarnos and Nassers and U Nus, or whatever their names may be. But even in these areas a cooling off may be expected in the near future. This is certainly a gratifying change in the climate of economic-policy discussions. However, a good many of the things which the open or—still more—the secret enemies of the market economy now demand under new labels come perilously close to those which are so discredited by their old name that one prefers not to mention them.

Nevertheless, it would be a mistake to overestimate the victory of the market economy or to think that its fruits are secure. First of all, we must not forget that the victory is anything but complete. We may perhaps neglect, in this context, the fact that the domination of total collectivism over approximately one-third of mankind remains unbroken, notwithstanding some minor concessions to personal independence and decentralization. But a good many countries of the free world, too, are still permeated by considerable remnants of collectivist policies, the elimination of which meets with stubborn resistance, even under non-socialist governments like the British. Some countries, especially in Scandinavia, are still so

26

strongly influenced by the principles, institutions, and ideologies of the socialist welfare state that their efforts to combat the ensuing inflationary pressure do not hold out much promise of success. This is one of the reasons why the re-establishment of complete currency convertibility on the basis of balanced international economic relations remains an unsolved problem in spite of some progress and occasional determined efforts. It follows that the real basis of international economic integration also remains incomplete, that is, a universal currency system with free and stable foreign-exchange markets. This, in turn, creates a fertile field for all kinds of attempts at international economic controls. Certain aspects of European economic integration show quite clearly how this situation reacts on the separate national economies and weakens the market economy.

Furthermore, we would be greatly deluding ourselves if we regarded the market economy as secure, even in countries like Germany. First of all, we have to remove any misconception about what really happened. The German economic reform was not simply a once-and-for-all act of liberation, a removal of obstacles that blocked the way to an automatic and natural process of recovery and growth. It was not like that at all. The history of German economic policy since 1948 has proved that economic freedom is like any other freedom: it must, as Goethe says, be conquered anew each day. The act of liberation was a necessary but not a sufficient condition for recovery and growth. The German example shows the market economy must be won and secured over and over again and not just against new dangers and temptations or in the face of new tasks. Since 1948, the German market economy has had to grapple continuously, both with old and persistent problems and with new and changing ones. It was now this and now that, now foreign trade and now the capital market, the budget, social tensions, agriculture, or transport. Residues of collectivism, such as rent control, were scattered about the market economy like unexploded mines, and they proved to be exceedingly difficult to dispose of through normal democratic procedures and under the cross fire of vote-catching demagogy.

On all of these and other fronts, German economic policy has continued to fight, with varying fortunes. Some grave and almost ir- reparable mistakes were made. Some brilliant successes were scored, especially in the revival and expansion of foreign trade and the rehabilitation of the currency. In between the positive and negative poles there is a whole gamut of more or less satisfactory average results, half or three-quarter successes and—as in the case of agri- culture and public finance—hitherto unavailing attempts to solve intractable chronic problems. The most serious feature was, and to some extent still is, capital shortage; in spite of all the progress achieved and in spite of an enormous rate of investment, this re- mained for a long time a dead weight on the German market econ- omy, and its effects were aggravated until quite recently by the failure to create a really free and well-functioning capital market. There can be no doubt that in this field the German market economy suffered one of its most conspicuous defeats, and even though the setback was due to the disregard rather than to the application of the market economy's own principles, the fact remains that there was a serious threat from this quarter. The situation is much the same in Austria. Indeed, it is no exaggeration to say that nowadays the central and most pressing problem everywhere is how to insure the continuing economic growth of the countries of the free world by means of adequate capital supply resting on true saving and not on inflation and taxation or too much on business profits (self- financing).

Let us leave the instructive example of Germany and return to the general prospects of the market economy in its contest with collectivism. It can, I think, by no means be taken for granted that the nature, conditions, and operation of the market economy are generally understood—in spite of all the lessons of experience and the best efforts of economists. Otherwise, how could people again seriously moot the idea of containing inflationary price rises with a new set of price ceilings and controls, as if the memory of decades of repressed inflation had simply been blotted out? Thus it is that the field still remains open, incredibly enough, to planned-economy sallies in various guises and in various places.

People are still in the habit of taking refuge in official regulations whenever a new problem turns up. In Europe, this takes the particularly absurd form of expecting any problem found intractable on the national scale to be solved on an international scale by a supra-national authority. Behind the façade of the market economy, people are still, consciously or unconsciously, promoting a development which leads to bureaucratic rigidity and the omnipotence of the state. They still tend, in the name of economic and social security, to heap new tasks on the government and thereby new burdens on the taxpayer.

Again and again we see ourselves cheated of the hope of reducing to tolerable limits the crushing weight of taxation, which in the long run is incompatible with a free or even moderately sound economy and society. It is not unusual today for the government's budget to absorb as much as 30 or 40 per cent of the national income through various kinds of compulsory contributions. This reinforces inflationary pressure and has a disintegrating and ultimately paralyzing effect on the market economy. In the presence of such excessive fiscal burdens, the market mechanism no longer works in the manner which theory assumes and economic order requires. The whole process of the economy is distorted by those households and firms whose financial decisions are made with an eye toward the tax collector rather than toward the market, and incentive is weakened at all levels and in all spheres. The tax system, which today is highly intricate and impenetrable but is at the same time of decisive importance for individuals and firms alike, has become, in the hands of government, a no less insidious than effective tool with which to sway and distort the market economy's processes and the natural selection of firms according to their true

29

market performance. Saving is depressed below the level necessary to finance growth investment without inflationary credit expansion, and at the same time, the rate of interest loses its essential efficacy because, as a cost factor, it takes second place after tax payments. Thus the capital market is upset, and ominous encouragement is given to inflationary tendencies. Old-style public finance is turned into a kind of fiscal socialism which more and more socializes the use of income. It is, unfortunately, becoming quite plain that the combination of the overexpansion of public expenditure and the reorientation of its purposes in socialist directions constitutes a source of continuous inflationary pressure and is incompatible with the market system in the long run.[9]

Still another circumstance accentuates this development. Not satisfied with their already enormous holdings of publicly operated enterprises, most governments, spurred on by the vested interests of the civil service, are still trying to acquire more and more such holdings and thus create veritable strongholds of public power and monopoly. This is happening even in Germany, the model of the market economy, to say nothing of Italy or France. At the same time, there remains a sense of social grievance, a hostile and economically irrational distrust of everything that goes by the name of capital or entrepreneur, together with a stubborn misconception of the latter's task and the conditions in which he can fulfill his functions, which are essential to the market economy. Free economy stands or falls with the free entrepreneur and merchant, just as such an economy is inconceivable without free prices and markets. There is no way of defending the free economy against the still powerful forces of collectivism except by having the courage to stand by these central figures of a free economy and protect them from the wave of distrust and resentment to which—more in the Old World than in the New—they are exposed.

We can do this more confidently and effectively if more entrepreneurs embrace free competition, which makes them the servants of the market and causes their private success to depend upon their services to the community. Otherwise they stab us in the back. But the task of safeguarding free competition and preventing concentration of economic power is exceedingly difficult and at best cannot be solved without compromises and concessions. At the same time, of course, there is the related task of making sure that competition does not degenerate in any way but remains a fair fight, so that the only road to business success is through the narrow gate of better performance in the service of the consumer and not through the many back doors of unfair and subversive competition, which are only too well known to the business world. In fact, these twin tasks have so far not been solved even tolerably satisfactorily in any country. At best they are being seriously tackled, as in the United States and Germany; at the worst no notice is taken of them at all.

Nevertheless, the progress made in safeguarding free competition among producers and protecting it from economic domination is sufficient to hope for eventual full success. But another monopoly position is gaining ground with uncanny speed and has, for very profound reasons, developed into the strongest and most dangerous bastion of social and economic power, and that is the monopoly of labor unions. This monopoly position remains unassailed, if indeed it is not further waxing in strength and danger. The concentration of supply on the labor markets in centralized trade-unions works with the whole arsenal of monopoly power and does not shrink from naked extortion. This monopoly is the most damaging of all because of its all-pervading effects, the most fatal of which is the inflationary pressure of our times. And yet the nature and significance of this labor monopoly are recognized by only a few, and even for them silence is the counsel of wisdom, unless they are free and independent or have the courage to face the consequences of openly stating unpopular truths. But since there is only a handful of people in the modern world with enough freedom and independence to save themselves from the suicidal effects of such courage, it is easy to imagine the prospects of solving a problem which cannot even be raised.[10]

31

Behind all of these perils we always encounter one predominant problem which we must face whenever we stop to think about the fate awaiting the industrial nations that are built upon the principle of economic freedom, a fate which these nations approach with alarming complacency or even with pride in something they call progress. This all-pervading problem is the process of growing concentration in the widest sense and in all spheres: concentration of the power of government and administration; concentration of economic and social power beside and under the state; concentration of decision and responsibility, which thereby become more and more anonymous, unchallengeable, and inscrutable; and concentration of people in organizations, towns and industrial centers, and firms and factories. If we want to name a common denominator for the social disease of our times, then it is concentration, and collectivism and totalitarianism are merely the extreme and lethal stages of the disease.

We all know what consequences progressive concentration entails for the health, happiness, freedom, and order of society. First of all it destroys the middle class properly so called, that is, an independent class possessed of small or moderate property and income, a sense of responsibility, and those civic virtues without which a free and well-ordered society cannot, in the long run, survive. The obverse of the same medal is the steady increase in the number of those who are not independent, the wage and salary earners, whose economic focus is not property but money income. The workers and employees are progressively merging into a uniform type of *dependent labor,* the teeming millions which populate the factories and offices of giant concerns. It may be that in many cases the large firm has a superiority of technical and organizational methods, although this superiority is often exaggerated and frequently rests merely on the artificial, though perhaps not deliberate, support of the government's economic and fiscal policy. But if this means denying man and the society determined by human values their due, then our accounts go seriously wrong, and this miscalculation may become the source of grave perils to free society and free economy.

However, the most immediate and tangible threat is the state itself. I want to repeat this because it cannot be stressed too much. The state and the concentration of its power, exemplified in the predominance of the budget, have become a cancerous growth gnawing at the freedom and order of society and economy. Surely, no one has any illusions about what it means when the modern state increasingly—and most eagerly before elections, when the voter's favor is at stake—assumes the task of handing out security, welfare, and assistance to all and sundry, favoring now this and now that group, and when people of all classes and at all levels, not excluding entrepreneurs, get into the habit of looking on the state as a kind of human Providence. Is it not precisely this function which increases the power of the state beyond all bounds, even this side of the Iron Curtain? Is Frédéric Bastiat's century-old malicious definition of the state as *une grande fiction à travers laquelle tout le monde s'efforce de vivre aux dépens de tout le monde* not now becoming an uncomfortably close fit?

The bloated colossus of the state, with its crushing taxation and boundless expenditure, is also chiefly to blame for the smoldering inflation that is a chronic evil of our times. Its destructive action will continue as long as the state does not radically curtail its program and as long as we do not drastically revise some of the most cherished popular ideas of present-day economic and social policy, such as full employment at any cost, the welfare state, the use of trade-union power for inflationary wage increases, and so on. But what hope of this can there be in a "dependent labor" society and a mass democracy afflicted by concentration?

It is evident that the actually existing forms of market economy, even in Germany, Switzerland, and the United States, are a far cry from the assumptions of theory. What we have is a hodgepodge system in which the basic substance of the market economy is not always easily discernible, a cacophony in which the dominant note

33

of economic freedom is not always clearly audible. The fact that the market economy still functions in spite of a hitherto unimagined degree of intervention is no proof that such distortions and handicaps are harmless, let alone useful. On the contrary, if it proves anything, it is the astonishing resilience of the market economy, which is obviously hard to kill.

On the other hand, there can be no doubt that there exists a critical point beyond which the symptoms of strain on such a market economy become alarming. This is a problem which has so far never been treated with the attention it deserves. One thing is certain: an excess of government intervention deflecting the market economy from the paths prescribed by competition and price mechanism, an accumulation of prohibitions and commands, the blunting of incentives, official price-fixing, and restrictions on primary economic freedom must lead to mistakes, bottlenecks, less-than-optimal performances, and imbalances of all kinds. At first these may be overcome comparatively easily, but with the proliferation of intervention, they end up in general chaos. The worst is that the disturbances caused by intervention are often taken as proof of the inadequacy of the market economy and so become a pretext for more and stronger intervention. It needs more understanding than can be generally expected to appreciate that intervention is at fault. Rent control, which, as every well-informed person knows, outdoes everything else in injustice and economic irrationality and not only tends to perpetuate the housing shortage but places an additional burden on the capital market, is a glaring and most depressing example, even in the model countries of the market economy.

This survey of the perils which today surround the market economy has shown that it is not nearly in such good condition as its outward success might suggest. We are concerned about the market economy. It is not that we think it is wrong; on the contrary, the reasons for defending it are as compelling as ever. It is precisely because we know how infinitely important it is to maintain, protect, and develop it in the face of the collectivist menace that we fear for the market economy in a world where social and political conditions are, on the whole, against it and threaten to become worse unless we remain vigilant and active. Market economy, price mechanism, and competition are fine, but they are not enough. They may be associated with a sound or an unsound structure of society. But whether society is sound or unsound will eventually decide not only society's own measure of happiness, well-being, and freedom but also the fate of the free market economy. Market economy is one thing in a society where atomization, mass, proletarianization, and concentration rule; it is quite another in a society approaching anything like the "natural order" which I have described in some detail in my earlier book *Mass und Mitte*. In such a society, wealth would be widely dispersed; people's lives would have solid foundations; genuine communities, from the family upward, would form a background of moral support for the individual; there would be counterweights to competition and the mechanical operation of prices; people would have roots and would not be adrift in life without anchor; there would be a broad belt of an independent middle class, a healthy balance between town and country, industry and agriculture.

The decision on the ultimate destiny of the market economy, with its admirable mechanism of supply and demand, lies, in other words, beyond supply and demand.

CHAPTER II

Modern Mass Society

Some twenty years ago, I made a first attempt to describe the nature of modern mass society and used the term *Vermassung*, "enmassment," to characterize it. I also tried to point to the extraordinary dangers threatening every aspect of our civilization, not least the economic and social. I tried to demonstrate that the danger was immediate and that it was continually becoming more evident and more difficult to avert, while the world, in the midst of all its disasters, still clung to the consoling idea of material progress, sometimes to the exclusion of all else. So much has happened in these twenty years that I feel I must try again, perhaps more insistently, and certainly in the light of new purposes and new events. While the discussion of this subject has in the meantime grown to vast dimensions, the features of modern mass society have become sharper and more unmistakable and its dangers, on the whole, more appalling and alarming,[1] certain equilibrating, healing, and compensating processes notwithstanding. This makes my task a good deal easier. Today I can rely on my readers' being familiar with the nature and the problems of mass society and of the process of enmassment which generates it. Rather than repeat matters which have long ago been stated and understood, I can turn to the less thankless task of saying what seems to me most important at this juncture. Taking advantage of the presumed general understanding

36

of the issue, I feel that I need not worry too much about presenting a coherent and complete picture and can safely leave my readers to fill in the gaps from their own knowledge and experience.

The first difficulty crops up at once. The pioneers who early recognized and diagnosed the symptoms of modern mass society— one of the first and most distinguished of whom, Ortega y Gasset, recently left us in old age—must today be inclined to observe somewhat wistfully that their endeavor has been overtaken by the common fate of new ideas. It is a fate which in itself exemplifies one of the laws of mass society: what was once fresh and meaningful becomes somewhat like a coin which passes from hand to hand and loses its sharp outline in the process.

The more "enmassment" has been talked about in the last fifteen years, the more shallow and blurred the concept has become. At the same time, a certain amount of exaggeration leads us to see "mass phenomena" everywhere. Occasionally, one meets with a tendency to use such words somewhat smugly and self-righteously— as if "mass man" were always the other fellow—as an expression of some sort of vague uneasiness or of genuinely reactionary dissociation from everything which may be called "the people." The very success of a new watchword is often its undoing, because it is hard to hear it up and down the country without eventually coming to regard it as hackneyed and then to react successively with weariness, criticism, and, finally, antagonism. This is probably the most charitable explanation for the fact that some smart alecks nowadays pretend that all the fuss about mass and enmassment is a false alarm and that the disintegration of society described by these words is only a new, and anything but pathological, stage of cultural development. These people would have us believe that everything is as it should be and that paradise is just around the corner: the paradise of a society whose idea of bliss is leisure, gadgets, and continuous fast displacement on concrete highways.

There are those who lightheartedly dismiss this great anxiety as intellectual twaddle and unashamedly talk about a new cultural

37

style just because of the ever rising consumption of things by which
the standard of life is thoughtlessly measured. Some of these people, I say, behave like the students in Auerbach's cellar:

> As 'twere five hundred hogs, we feel
> so cannibalic jolly.

But Mephistopheles' answer is not far away:

> Not if we had them by the neck, I vow,
> Would e'er these people scent the Devil!

What these happy-go-lucky children of our time do not seem to
grasp is that to regard the problem of modern mass society as an
invention of intellectual dreamers is to misunderstand the central
issue of our epoch. It is the crucial issue for the moral, spiritual,
political, economic, and social future of the world into which we
are born. Whoever denies this ought to reflect that in ignoring or
belittling the enmassment problem he is at one with the Communists and so furnishes one more proof of the close affinity between
a certain brand of Western thought and Communism. "Mass," to
take an example at random, is defined in the Communist *Lexikon
A-Z in einem Band* (Leipzig, 1955) only as "the masses of the
workers as a whole," and Ortega y Gasset is described as an "extreme reactionary and individualist philosopher."

The very mildest answer which such glossing over and extenuation calls for is that mass society is a phenomenon for which we
lack historical standards and one for which we are unprepared by
the experience of earlier ages. This is certainly true of the all-
pervasive character of mass society and of its foundations, if not,
perhaps, of all its separate elements. This phenomenon accounts for
a large part of the revolutionary change which has taken place
during the last forty years in the manner of human life and
thought and which radically distinguishes our epoch from all past
ones. This revolution has occurred within the lifetime of the more
elderly amongst us, including myself. Together with the triumphs of

technology, urbanization, and industry and with the still rising flood of world population, the genesis and spread of mass society is one of the most important facets of this revolution.

Mass and World Population

Each of us brings his personal experience to the understanding of the problem under discussion. What the words mass society first call to mind is the visible crowdedness of our existence, which seems to get irresistibly worse every day: sheer oppressive quantity, as such, surrounding us everywhere; masses of people who are all more or less the same—or who are at least assimilated in appearance and behavior; overwhelming quantities of man-made things everywhere, the traces of people, their organizations, their claims. Merely to get along with all of these quantities requires constant adjustment, accommodation, self-control, conscious and practiced responses, and almost military uniformity. In the great cities of the United States, it is considered necessary that school children, instead of being taught more important things, should have lessons in "social adjustment," that is, in the art of queueing patiently, folding one's newspaper in the subway without being a nuisance to other passengers, and other such tricks of civilization. Even in Europe it would be interesting to investigate how much less history or other culturally important things children learn at school because they have to be taught the traffic code.

Nor is it easy nowadays to escape from the crowdedness of life and the flood of people in order to be alone for a while. In my experience, a typical Sunday outing in New York means taking one's car and driving out of town along an exactly prescribed traffic lane and in the midst of other endless columns of motorcars, stopping at the edge of a wooded area and parking the car in the space provided, paying one's entrance fee and looking, with thousands of others, for a free square yard to sit down on, stretching one's legs by walking up and down a few paces, and then returning, in the same manner as was used in coming out, to one's flat on the fifteenth floor of an apartment building. After that, one is supposed to be refreshed and ready to face the week's subway trips and office work somewhere up on the fiftieth floor, and perhaps might even be energetic enough to spend an evening at the movies in Rockefeller Center, in company with ten thousand others, having first, as part of an endless human serpent, shuffled through a cafeteria in order to stoke up with the necessary calories and vitamins.

To try to escape from the giant honeycombs of city dwellings into the suburbs is to jump from the frying pan into the fire—with reference, again, to some parts of the United States. Suburbia has its own very charming brands of mass living. The price of having a little house and garden of one's own may be a gregariousness that surpasses anything known in the center of town. There can be no question of one's home being one's castle. Everybody is forever "dropping in" on everybody else; the agglomeration of people stifles all expression of individuality, any attempt at keeping to oneself; every aspect of life is centrally ruled. When I was living on Long Island, even the temperature of the central heating—always excessive for my taste—was controlled at the common boiler-house for all the tens of thousands huddled together in this settlement. "To hoard possessions is frowned upon," writes American sociologist William H. Whyte, Jr., in his "Individualism in Suburbia" (*Confluence* [September, 1954], p. 321). "Books, silverware, and tea services are constantly rotated, and the children feel free to use each other's bicycles and toys without bothering to ask." Everyone is under pressure to join in, to take part in communal life, even if it means giving up or neglecting his private occupations, unless he wants to be known as a spoilsport. Classes on "family group living" end up by being more important than natural and free family life itself. Yet the development of a natural community is impossible, if only because of the constant coming and going of people.

We all know to what extent this American pattern of life has already spread to Europe. We can hardly hope to escape the same hell of congestion. In Europe, too, the traffic columns are becoming denser, and even the queues at ski lifts are getting longer; the stony ranges of our cities grow upwards and sideways; the very mountain peaks, which Providence seems to have preserved as a last refuge of solitude, are drawn into mass civilization by chairlifts. In Europe, too, the power shovels of the world of steel and concrete are advancing steadily.

As we increasingly become mere passively activated mass particles or social molecules, all poetry and dignity, and with them the very spice of life and its human content, go out of life. Even the dramatic episodes of existence—birth, sickness, and death— take place in collectivized institutions. Our hospitals are medical factories, with division of labor between all sorts of health mechanics and technicians dealing with the body. People live in mass quarters, superimposed upon each other vertically and extending horizontally as far as the eye can see; they work in mass factories or offices in hierarchical subordination; they spend their Sundays and vacations in masses, flood the universities, lecture halls, and laboratories in masses, read books and newspapers printed in millions and of a level that usually corresponds to these mass sales, are assailed at every turn by the same billboards, submit, with millions of others, to the same movie, radio, and television programs, get caught up in some mass organization or other, flock in hundreds of thousands as thrilled spectators to the same sports stadiums. Only the churches are empty, almost a refuge of solitude. Whether we travel or stay at home in our sprawling cities—and more and more of us are at home in them—it is becoming ever harder to escape the rising flood of people which drags us down and makes us creatures of the herd or the mass machinery which canalizes this flood. More and more we are becoming a part of this human compound, and I imagine that few of us can still harbor any doubts about what this means for man's spiritual and moral existence or for the health of society as a whole. Nor can there be any doubt about the

41

immediate cause of this deluge of sheer human quantity and all its consequences. The immediate cause lies in a crudely simple fact: the irresistible, overflowing *growth of population*. Before the Second World War, it was generally assumed that population growth would steadily slow down and eventually stop, at least in the most developed countries, but this prediction has been belied by the facts. The human flood has gone on rising with unbelievable speed throughout the world. No one can now speak of the possibility, let alone the probability, that the curve may flatten out in the foreseeable future. World population is growing at a rate of something like 1.5 per cent each year, which means that it doubles every fifty years. The flood is most spectacular just where the limits of capacity have very nearly been reached: in the populous countries of Asia— in Japan, India, Indonesia, and Burma. Especially impressive is the case of Egypt. This oasis, hemmed in on all sides by the desert or the sea and even now settled and cultivated to its utmost capacity, provides an alarming, concrete demonstration of what overpopulation in the stark literal sense of the word can mean. What is new and disturbing in the situation is that, contrary to so many forecasts, the earth's population is rising and promises to go on rising at a faster rate than before, even in some of the old industrial countries of Europe and in the United States.

As alarming as the population growth is the blindness with which its dangers are denied or simply overlooked. While Jules Romains rightly regards the world-population problem as *problème numéro un* (in his book of that title), very few people understand, let alone accept, this view. There is an important and interesting task here for psychologists, who might try to analyze and explain this *delirium mundi pullulantis,* with its criminal optimism, as Schopenhauer might have called it, its cult of quantity, its taboos, and its strange mixture of statistics and lullabies, not forgetting the extraordinary part played by demographic nationalism—for it is always the growth of one's own people, and never that of others, which gives cause for rejoicing. However, there is a difference of degree between the Western world and some of the Asian countries about to be drowned by the population flood. Under the pressure of compelling facts, public opinion in Japan, for instance, has by now come around to the view that the birth rate must be adapted to the death rate. The same attitude is gaining ground in India. But in the industrial countries of the West, large families are still regarded as a merit, to be rewarded by the treasury out of penalties on less procreative citizens, and it counts as a duty to the community to fulfill one's biological functions. Special emphasis should be given to the fact that this attitude, almost an "official" one in some countries, can no longer invoke the unequivocal dogma of the Catholic church, even though the latter has remained conservative on this subject for reasons which deserve respect.[2]

Thus the modern world affords the unique spectacle of people being as busy multiplying as in preparing their own extermination, and by the very same technology without which the demographic growth could never have occurred in the first place. Now one would expect that if people watch this expansion with so much optimism they must have some sound reasons for it, especially since this optimism frequently takes the form of almost arrogant self-assurance, occasionally combined with open contempt for the pessimists. However, the reasons are anything but convincing; they are not even plausible.

We are told that by improving agricultural methods and putting new land under cultivation we can keep pace with population growth for a long time to come. But such assurances ring false, even at present, especially if we are honest enough to include in the calculation such extremely important conditions of equilibrium between population and economic potential as raw material and power resources, as well as the symptoms of soil exhaustion and hydrological imbalance. Even now a country like India, in trying to raise the miserable standard of living of its expanding population, has to keep running in order to stay in the same place—unless capital imports were to increase far in excess of any practical pos-

43

sibilities. In this field, as in all others, it is incontrovertibly true that even the most prodigious investment (such as the much-discussed Aswan Dam in Egypt) provides only a more or less extended breathing space—until the hard-won elbow room is populated by new millions.

The race between technology and population growth is becoming ever more arduous, and the necessary capital expenditure implies an ever more stringent restriction of current consumption. Nor can the race be kept up forever. This must surely be obvious to anyone who stops to think. Even the most stubborn optimism will eventually have to yield to the irrefutable fact that this historically unique and explosive population growth simply must cease, unless the world, with its population expanding at the present rate, is to have three hundred billion inhabitants in a few hundred years (more precisely, in the year 2300). If it has to stop eventually, why not now? Why must the earth first be transformed into an anthill? Why must nature be completely ravaged, with all the concomitant risks for man, the "parasite of the soil," to use Edward Hyams' phrase? Must we really try to put off the inevitable at the cost of making hell out of our civilization? I should say that this would be nothing less than criminal irresponsibility towards the coming generations and a cowardly postponement of a decision which will have to be made in the end.[3]

This is the world population position. The industrial nations of the West, riding on the crest of a wave of economic expansion and mass prosperity, would be deluding themselves if they thought that they could remain safe from these long-run prospects for any length of time. Nevertheless, they are in a favored position. It does seem as if ever more intensive industrialization, along with the possibilities not yet fully appreciable that will be created by atomic energy and automation, has provided a formula for reconciling a growing population with a rising standard of living. In fact, certain bold theories would have us believe that it is precisely population growth, with ensuing mass demand and mass production, which imparts dynamism to the industrial countries.

I shall not for the moment go into the question of whether this formula, even at present, achieves all that is claimed for it. The problem of combining population growth with rising standards of living requires far more precise and profound research than is usually accorded it. In some respects we may well be better off than the preceding generation, but in other respects we are a good deal poorer.[4] But even if there were less doubt about all this than there is, it would be rash to overlook the exorbitant and still rising price which the Western industrial nations have to pay for this formula: the increasing vulnerability, instability, uncertainty, and susceptibility of an economy running in top gear.

The economic system of these countries, together with the social system which surrounds and supports it, can be likened to a pyramid standing on its point. It is the work of man, more ingenious than anything imaginable but also more artificial, intricate, and, in the aggregate, more vulnerable to any failure of one of its parts, which failure a simpler and more robust system could survive without harm. The welfare and existence of millions of people depend upon the orderly functioning of this huge mechanism, but with their mass passions, mass claims, and mass opinions, these same people are undermining the conditions of order, certainty, and sober reason, without which the greatest technical and organizational progress is of no avail.

Thus the foundations of a civilization to which we entrust the growing welfare of a rising population are weakened rather than strengthened by mass society, which is at the same time the result of this civilization and the origin of the population growth. We must ask ourselves whether, in the long run, the Western countries' mass supply system, with all its imponderable conditions, is at all compatible with the apparently irresistible development by which wage and salary earners are becoming the overwhelming majority of the population everywhere. If we have to buy their good graces by continually yielding to economically irrational demands and by the continuous expansion of a welfare state which stifles responsibility, incentives, and initiative alike and if we have to pay for this with chronic inflation, for how long can we be at all confident that technical progress will continue to yield the overall utility expected of it?

These are not idle questions, and no one who looks about with open eyes will regard them as such. But the mere circumstance that they can be seriously raised and that they admit, at least, of the possibility of a negative answer should be enough to shake the optimists' faith in an assured future for the world's teeming millions. The argument should be clinched, surely, by the consideration that the tenderest spot of our system is its dependence on orderly international economic relations, a dependence inadequately measured by the mere statistical share of the industrial nations' foreign trade in their national product. International economic relations can be reasonably secure only if they form part of a political and ethical international order. We once had such an international order, but today, when the industrial nations of the West are improvidently taking a new demographic upsurge for granted, it has become a tragic certainty that this old international order—doomed, perhaps, by some ineluctable fatality—has broken down without being replaced by another.[5] We had the first inkling of what this signifies when we found that today the whim of an Oriental despot is enough to paralyze the proud millions of our motorcars and to turn our winter heating into a problem.

These sober reflections may cause some expansionists to cling all the more firmly to the above-mentioned theory, which regards demographic growth in the industrial countries of the West as an indispensable motive power of the economy and as self-supporting, thanks to the bounty of an economic system so dynamized. This sounds like the tale of a modern Baron Münchhausen who is trying

46

to pull himself up by his own bootstraps. Everything goes like clockwork: the more cradles there are in use, the greater is the demand for goods, the higher is investment, the fuller is full employment, the more vigorous is the boom, and the more dynamic the economy.

This theory is astonishingly popular, but it is highly dubious, to say the least, and in this simple form certainly wrong. The mere fact that until recently population growth was, on the contrary, regarded as a source of unemployment should make us suspicious. The example of the populous underdeveloped countries and, in Europe, of Italy proves that this can indeed be true in certain circumstances, namely, when population growth is associated with lack of capital, so that it is difficult to absorb the additional labor force without a decrease in wages. Conversely, we have to admit that population growth may help to overcome economic depression in conditions of excess capital caused by insufficient investment of savings. However, it remains doubtful whether such "structural" support of the capital market by means of demographic expansion continuing through cyclical fluctuations is a really effective counterweight against the psychological and monetary factors causing these fluctuations. A still more pertinent question is whether, in the presence of an intelligent countercyclical policy, the remaining stimulating forces of the modern economy would not be entirely sufficient.

In any case, these considerations are not topical today anywhere in the Western world, since it suffers not from a surplus but from a deficiency of capital and from inflationary excess investment caused in large part precisely by the rising populations' additional capital demand for housing, means of production, roads, and many other things. In these circumstances there is a presumption that without the stimulant of demographic expansion, we would get, instead of depression, a salutary and anti-inflationary limitation of investment to projects designed to raise the productivity and welfare of a stable population. No convincing reason compels us to expect any decisive change in this situation in the foreseeable future.[6]

For the rest, the champions of the theory here refuted generally do not seem to realize that it is a degradation of man and of the great mystery of creation to turn conception and birth into an indispensable means of raising the demand for motorcars, refrigerators, and television sets—a mere mathematical factor, as it were, in the production-consumption equation. If these people were right, it would be a most striking demonstration of how far we have already traveled along this demographic path leading to a life ever more artificial, contrived, and unstable.

Nor is this all. There is one more argument, a decisive one, which brings us back to the beginning of these reflections. Suppose that everything said so far is less convincing than I believe it to be. Suppose that technology, science, inventions, and organization can really keep in step with population growth or even get ahead of it. Suppose that we can somehow cope with the problem of the exhaustion of the soil and of known raw-material resources. And suppose that there occurred nothing but the gradual transformation of the world into some sort of colossal urban complex, broken up by sparse green patches, rather like the Ruhr today. With all of these implausibly favorable assumptions, what would we get, beyond the material fact that people are reasonably well off, have enough to eat, and possibly consume a growing quantity of phonograph records and automobile tires? For how much does that count in the immeasurable non-material and therefore infinitely more important sense? In other words, what happens to man and his soul? What happens to the things which cannot be produced or expressed in monetary terms and bought but which are the ultimate conditions of man's happiness and of the fullness and dignity of his life?

This is the question. The fact that it is hardly ever raised constitutes one of the bitterest commentaries that can be made on our times. Is it not, we may modestly ask, part of the standard of living that people should feel well and happy and should not lack what Burke calls the "unbought graces of life"—nature, privacy, beauty, dignity, birds and woods and fields and flowers, repose and true leisure, as distinct from that break in the rush which is called "spare time" and has to be filled by some hectic activity? All these are things, in fact, of which man is progressively deprived at startling speed by a mass society constantly swollen by new human floods.

The optimists, with their illusions about population growth, are wrong at all levels of reasoning, and they are more in error at each successive level than at the preceding one. In the first place, they miscalculate with respect to the purely *economic optimum* size of population. They are defeated at the second level, where reflection on the *social optimum* of population size raises the question of whether demographic expansion beyond a certain point does not endanger the stability of the economy's social framework. And they are in an altogether hopeless position at the third, last, and most important level, where the *vital optimum* is at stake. There are sound reasons for assuming the social optimum to lie well below the economic, since the social and moral instability of the system can assume alarming proportions, although technical and organizational progress does not yet seem to give grounds for disquiet, but it is obvious that the vital optimum lies very much lower still. There is little doubt that the industrial countries of the West, let alone the nations of Asia and Africa that are vegetating on the border line of absolute overpopulation, have already gone beyond both the economic and the social optimum, and it is quite certain that they have left the vital optimum far behind and are rapidly approaching the vital maximum.

Since these things can be neither measured nor weighed, the vital population optimum is obviously not susceptible of scientific definition and determination by the usual methods of sociology and economics. But this does not disprove the paramount importance of the problem, though it challenges the methods of the social sciences, which have yet to learn that precision and truth are two very different matters. The social sciences lose all their meaning when their equations and statistics make them lose sight of the typical office worker in New York, who has to pay the price of human agglomeration by having to spend three hours a day commuting between his home and his office, who jumps from one means of transport onto another and is lucky if he finds a seat—provided, of course, that he does not set himself up as an outmoded, quixotic knight of chivalry in our era of equality between the sexes. These conditions clearly figure as a liability in the balance sheet of his life. It will be small consolation to him that the item, counted as part of the transport component of national income, reappears as an asset in standard-of-living statistics. All this does is to illustrate the absurdity of such a standard-of-living philosophy.

If we are to estimate correctly the vital significance of the present-day flood of humanity, we must not forget the New York commuter, so typical of our era. But neither must we forget all the other consequences of that flood which are manifest in the congestion of our existence. We must not forget the progressive disappearance of the difference between town and country, glorified as an ideal by Marx and now taken for granted even by non-Marxist sociologists, with a disdainful smile at our backwardness; or the noise and stench of mechanized mass living; or polluted rivers in which we can no longer bathe; or the increasing difficulty of assuring an adequate supply of drinking water; or the horrible violation of nature, which we are gradually turning into a desert and the balance of which we disturb and finally destroy, to our own incalculable damage. And finally, returning to the main stream of our argument, we must not forget mass society as a whole, whose primary origin or condition is the human flood itself.

It puts our time and its sociologists to shame that a man like John Stuart Mill, one of the fathers of nineteenth-century utilitarianism, clearly saw, more than a hundred years ago, the issue which we characterize as the problem of the vital population optimum. In his *Principles of Political Economy* (Book IV, Chapter 6, §2), which, like the *Communist Manifesto*, appeared in 1848, we find the following:

"There is room in the world, no doubt, and even in old countries, for a great increase of population, supposing the arts of life to go on improving, and capital to increase. But even if innocuous, I confess I see very little reason for desiring it. The density of population necessary to enable mankind to obtain, in the greatest degree, all the advantages both of co-operation and of social intercourse, has, in all the most populous countries, been attained. A population may be too crowded, though all be amply supplied with food and raiment. It is not good for man to be kept perforce at all times in the presence of his species. A world from which solitude is extirpated, is a very poor ideal. Solitude, in the sense of being often alone, is essential to any depth of meditation or of character; and solitude in the presence of natural beauty and grandeur, is the cradle of thoughts and aspirations which are not only good for the individual, but which society could ill do without. Nor is there much satisfaction in contemplating the world with nothing left to the spontaneous activity of nature; with every rood of land brought into cultivation, which is capable of growing food for human beings; every flowery waste or natural pasture ploughed up, all quadrupeds or birds which are not domesticated for man's use exterminated as his rivals for food, every hedgerow or superfluous tree rooted out, and scarcely a place left where a wild shrub or flower could grow without being eradicated as a weed in the name of improved agriculture. If the earth must lose that great portion of its pleasantness which it owes to things that the unlimited increase of wealth and population would extirpate from it, for the mere purpose of enabling it to support a larger, but not better or a happier population, I sincerely hope, for the sake of posterity, that they will be content to be stationary, long before necessity compels them to it."

Wise words—and unheeded ones. More than a century later, one of the leading journalists of the United States, Edgar Ansel Mowrer, in an article in *The Saturday Review* (December 8, 1956) entitled "Sawdust, Seaweed, and Synthetics," accused the demographers of "appalling inhumanity" in solemnly calculating the "amount of nourishment necessary to fill ever more billions of bellies" without stopping to ask themselves whether life will still be worth living when we have to rely on sawdust, seaweed, and synthetics. "Personal freedom, nature, beauty, privacy, solitude, variety, savor—trivial?" Even a positivist sociologist would have to admit that these are decisive values. "How long could a rashly multiplying mankind continue to find enough beauty?" Mowrer points out that if population continues to increase, the authorities would necessarily have to curtail our liberties more and more, "not because they are necessarily opposed to freedom but because they must do so if living is to be made endurable for any of [the people]." A hundred years ago, Mill had no doubts on this subject. But perhaps his dark forebodings had first to come true, so that in our own days a man like this American journalist could ask the soul-searching question: "How far can mankind lose contact with both the organic substratum and the macrocosmic framework of his life and prosper spiritually?" Just how narrow the margin is, in fact, we shall see presently when we discuss boredom as the curse of our modern mass society.

Mass—Acute and Chronic

We must probe deeply in order to understand the true nature of mass society. The everyday experience of crowd and quantity, however elementary it may be, does not exhaust the problem. Behind it something else is happening, something much more profound, significant, and fateful: a shift of the center of gravity towards collectivity and away from the individual, at rest within himself and holding his own as an integral personality. The equilibrium between individual and society, their relation of constant tension and genuine antinomy, is disturbed in favor of society. This equilibrium—there can be no doubt whatever about it—is the norm of individual and social health. We do not hesitate, therefore, to call the serious imbalance a disease, a crisis, with which we cannot live for long.

In the most general terms, our conception of mass society is this: The individual is, in our time, losing his own features, soul, intrinsic worth, and personality because and in so far as he is immersed in the "mass," and the latter is "mass" because and in so far as it consists of such "depersonalized" individuals. To the extent of this shift of the center of gravity, the essential element which the individual needs in order to be a complete human being and spiritual and moral personality seems to us to be missing. At the same time, one of the basic conditions of sound social life is destroyed.

This is the fundamental idea common to all theories of the evils of "mass": As part of "mass," we are different from what we normally are and should be in healthy circumstances; we are subhuman, herdlike, and the state of society dangerously corresponds to our own. This much is common ground, but further analysis may proceed in a number of different directions, which should be sharply distinguished lest we aid and abet the blurring of the concept of mass. We must begin by distinguishing two kinds of mass in the sense of social pathology: mass as an acute state and mass as a chronic state.

Mass as an acute state can occur in any historical context whenever individuals temporarily become part of a crowd, an inorganic agglomeration which does not interfere with the individual's form of life and which dissolves as quickly as it is assembled. I have in mind such things as a mass meeting, a sudden mass movement, or even some spiritual epidemic. In such cases, individuals, as we know, are subject to certain psychological laws whose effect may be characterized as emotional hyperthymia, intellectual regression, and paralysis of the moral sense of responsibility. In other words, as parts of an acute mass, we are more excitable, more stupid, and more ruthless than usual. It is acute mass with which, in the main, the literature on mass psychology has been concerned from the time of Gustave Le Bon to our own day. Its findings and our corroborating experience are no doubt important and interesting, but this is not really the issue that concerns us when we speak of enmassment as a disease of modern society. It is true that enmassment fosters the emergence of acute mass by multiplying the occasions on which we are subject to the laws of mass psychology. It is equally true that acute mass may give rise to very perilous conditions so that we are well advised to guard against it and to arrange our public and social life in such a manner as to protect it, as far as possible, against the avalanches of mass psychology. But the phenomenon is neither novel—Peter of Amiens was very successful in bringing it about in the eleventh century—nor of such a menacing kind as to lead us to consider enmassment as one of the central problems of the present social and cultural crisis. The former problem is not even part of the latter, but of a different kind, which does not prevent the two from being constantly confused.

The problem with which we are actually faced is the problem of mass as a chronic state, as a permanent condition of the life of more and more people in the world today. Here again we have to distinguish a number of different aspects and facts.

The process may be understood in the first place as essentially belonging to the intellectual and ethical sphere. What we have in mind is the way thought is becoming shallow, uniform, derivative, herdlike, and tritely mediocre; the growing predominance of the semi-educated; the destruction of the necessary intellectual hierarchy of achievement and function; the crumbling away of the edifice of civilization; and the presumption with which this *homo insipiens gregarius* sets himself up as the norm and chokes everything that is finer or deeper. In this context, a highly significant detail to be noted is the fact that classical education seems to be doomed in our mass society, if only because mass man persecutes it with genuine hatred, the hatred of one whose lack of mental discipline bars him from access to this education—not to mention the influence of our era's rampant utilitarianism, technicism, and materialism.[7] It is a case of aggrieved educational Jacobinism, intellectual egalitarianism aiming at the leveling out of "mental income inequalities," not upwards, as would be desirable but is impossible, given the unalterable aristocracy of nature, but downwards, as is only too easily possible but quite definitely undesirable.

It is this process of intellectual and ethical depersonalization which Ortega y Gasset had principally in mind when he wrote his pioneer work *The Revolt of the Masses.* The crisis of our times is here seen chiefly as a cultural crisis. It cannot, of course, be isolated from the concurrent structural changes of society, but its real and deepest roots reach down to those ultimately decisive levels of the mind and the soul where the sustaining ideas and values of the Western world's Christian civilization are at stake. One would have to be very remote from these ideas and value judgments to fail to recognize enmassment, in this sense of intellectual and moral disintegration, as the real core of the evil or to deride such a diagnosis. Only if we overcome enmassment can we hope to save our seriously ailing civilization.

Now this process of everything becoming part of mass in the intellectual and moral sphere is supported by a similar process in the social sphere. By this we mean the disintegration of the social structure, generating a profound upheaval in the outward conditions of each individual's life, thought, and work. Independence is smothered; men are uprooted and taken out of the close-woven social texture in which they were secure (this is now happening on a large scale in the former colonial and dependent territories); true communities are broken up in favor of more universal but impersonal collectivities in which the individual is no longer a person in his own right; the inward, spontaneous social fabric is loosened in favor of mechanical, soulless organization, with its outward compulsion; all individuality is reduced to one plane of uniform normality; the area of individual action, decision, and responsibility shrinks in favor of collective planning and decision; the whole of life becomes uniform and standard mass life, ever more subject to party politics, "nationalization," and "socialization."

Urbanization, industrialization, and proletarianization (in a well-defined sense) are only special aspects of this sinister overall process, which I once tried to define metaphorically as a decomposition of the humus of society and its transformation into a social dust bowl. Now if mass thus becomes permanent, more and more people are in a constant state of readiness to submit to those psychological laws which the theory of acute mass describes. Advertising, propaganda, popular crazes meet no obstacles, since intellectual and moral enmassment have created a vacuum into which foul water can seep. Matters are made worse by the fact that mass persuasion now has at its disposal the new and incomparable media of radio, film, and, above all, television.

The danger of spiritual infection has, in brief, become enormous, especially since, in the sterile bustle of our times, more and more people (above all, those in responsible positions) lack the leisure and composure to think for themselves or to commune with the author of some thoughtful book. "Someone who is compelled every day to think of his colleagues, his parliamentary majority, his re-election, the press, public opinion, party intrigues, and a thousand other things ends up by being able to devote very little time to meditation, and even that must be reserved for problems important in relation to other people, problems of success," says Jules Romains. All the more enthusiastically do such men indulge in the conference mania, which gives them the illusion that wisdom is to be gathered at conventions or congresses and is a precipitate of the resolutions and terms of reference of committees and subcommittees. "It is hard to imagine a minister of Louis XV or, to take more familiar examples, a Talleyrand or a Metternich embracing some newfangled doctrine with so much enthusiasm that they lose their lucidity of mind," writes Romains. The more rapidly people in our

56

times succumb to spiritual mass epidemics and the less resistance their own judgment possesses, the more discontinuous and variable becomes the spiritual image of society.

If such people avidly lap up mass slogans, if they surrender to "social religions" as a surrogate for vanishing faith and traditional values, if they take to mass entertainment and mass spectacles almost as to narcotic drugs, they do it not merely to fill the emptiness of their souls. One of the principal reasons why they plunge into the mass is that they are made deeply unhappy by the social enmassment which prizes people out of the fabric of true community. Thrown into society as isolated human beings, literally *individua* and human atoms, people hunger for "integration," and they allay this hunger by means of the intoxicating thrills and crowds of mass society. They can no longer live without the radio, press, films, group outings, mass sports, and all the concomitant noise, without the sense of "being in the swim," without the "smell of the herd." We can count ourselves fortunate when these mass men rest content with this kind of thing and do not allow their hunger for faith and integration, their nostalgia for the consolation of firm inner and social support, to drive them into the maelstrom of the fearful social religions of our time—for these always imply fanatical and intolerant mass hatred (though frequently in the name of abstract, general philanthropy), such as national hatred, class hatred, and race hatred.

Mass Culture

Let us now look at the details of the overall process we have outlined. We begin with its intellectual and moral aspects. We are certainly not mistaken in regarding the civilization corresponding to mass society as mass culture or, as Guglielmo Ferrero called it nearly fifty years ago, "quantitative civilization." It is exemplified in today's mass products and in the mass tastes to which they appeal and of which they take commercial advantage. By whatever standards we measure cultural development, its curve has been going steadily downwards in all countries of the West during the last fifty years, notwithstanding certain brakes and compensations to which the optimists cling with a kind of despair. Men like John Stuart Mill, Herman Melville, or Jacob Burckhardt, who watched with misgiving the proliferation of the signs of cultural decay in their own times, would find their worst fears far exceeded if, halfway through our twentieth century, they could take the measure of the readers (or should I say the viewers?) of our illustrated papers and our race of educational pygmies. No doubt those who lay down the law in our society would ridicule them as hopelessly romantic because they have not yet grasped the fact that the hour of "modern consumption society" is at hand.[8]

Mass culture has gone quite far already, as can be seen by anyone who looks, without rose-colored glasses, at the indicators of intellectual mass consumption, from the seven-figure circulations of our completely infantile illustrated papers and the eighty to ninety million editions of the incredibly dreadful American comic books to the general educational and cultural level of our times. The reverse side of all this is the tight corner into which books really worth reading are driven, together with serious periodicals not catering to mass tastes. Nor is this all. It is hard to disagree with pessimists—such as, for example, the American Dwight MacDonald ("Mass Culture," *Diogenes* [1953], III)—who maintain that our civilization is becoming subject to a sort of Gresham's Law. Just as, according to Gresham's Law, bad money drives out good money, so, too, does modern mass culture make it increasingly difficult for anything better to hold its own. This situation, incidentally, also has economic reasons in the spread of commercialism and in the effects of mass-production methods on book and magazine publishing. Even where the finer product subsists, it is no longer the apex of a generally accepted pyramid, and there is a growing temptation for the intellectual elite to devote its talents to satisfying and flattering mass demand, thereby reaping both mass fame, however ephemeral, and substantial material gains. It is no use deluding ourselves. We are forced to admit that there is scant reason for the comforting conclusion that this process of decay has been arrested, let alone reversed. But there are enough encouraging symptoms on both sides of the Atlantic to render paralyzing despondency equally groundless.

No serious discussion of this subject can avoid a comparison between these results and the high hopes which a progress-happy era had pinned on the fight against illiteracy. We can but marvel that those who cherished these naïve hopes—some of them may still be about—never seem to have realized that what really counts is *what* all these people are to read once they have learned how to read. Nor do they seem to have asked themselves whether the standardized educational system by which illiteracy is eradicated was always favorable to a wise choice of reading matter. "The average Englishman reads nothing except a thin and vulgar daily newspaper, though he has been compelled to go to school for half a century; while in Portugal, the state with the highest rate of illiteracy in western Europe, the reading of serious books and journals, per head of population, is much higher than in enlightened Britain. The broad nineteenth-century public for English literature, in short, has very nearly ceased to exist." (Russell Kirk, *Beyond the Dreams of Avarice* [Chicago, 1956], 303–304)

That this should not be turned into a specious argument against compulsory education and the fight against illiteracy is so obvious that one is almost ashamed to mention the mere possibility of such abuse. But it is naïve to overlook the conditions on which depends the benefit of general education; these conditions are of more importance than teaching the technique of reading. It is mass society which has so largely destroyed these conditions.

Let me illustrate this point with an example from my personal experience. Not long ago I had occasion to discuss with a student the final draft of a paper; his nationality is immaterial. It was an above-average study of civil aviation, and the author was a mature,

59

experienced man whose education probably surpassed that of many of his kind. He had come to the conclusion that the economic rationality of this latest means of transport appeared somewhat doubtful if allowance were made for all the open and indirect subsidies. I brought our long conversation to an end with a few philosophical reflections and added that the story of Daedalus and Icarus still seemed to contain a mysterious truth. What kind of people were they, I was asked, and what in the world did they have to do with aviation? Did he not remember Ovid's *Metamorphoses?* No, that had never been mentioned in his Latin class. Had he never encountered this legend elsewhere? Again, no.[9]

Now let me tell another story, also a personal recollection. Many years ago I visited a second-hand bookshop in Istanbul. It was run by a Greek, and I found him immersed, together with a young girl, in the study of a book. I asked him not to let me disturb him, but while I was browsing among the dusty shelves, I could not help overhearing some of the remarks passing at the table. Soon there was no doubt: they were reading and discussing the *Odyssey.* It seemed to me that there could hardly be a more touching sight than that of this Greek, here, in a dark corner of ancient Byzantium, handing down to his daughter the eternal beauty of Homer, still a living heritage after three thousand years, while outside the trams rattled past and the motorcars hooted.

These two experiences, juxtaposed, illustrate the meaning of discontinuity and continuity in cultural tradition. They show what continuity signifies and, on the other hand, how sharp a break is taking place in our generation after a long process of attrition. It is a break which amounts to a cultural catastrophe, for we witness the passing of millennial traditions that have furnished the substructures of our civilization. Anyone who wants to instruct himself in detail about the sources, significance, and development of these traditions will find an admirable guide in Ernst Robert Curtius' great work *European Literature and the Latin Middle Ages.* Curtius has shown us on what an almost inexhaustible source we are still drawing, and in teaching us to appreciate it fully, he makes it all the more evident just what we are losing now, three thousand years after Homer, my amiable Greek in Istanbul notwithstanding.[10]

What were the instruments of this millennial tradition? Who were the people who took the biggest part in handing it down, and what economic and social conditions favored or disfavored the process, up to the threshold of our mass epoch? Who—this is the key question—read books in different periods of history and what books? It is not easy for us to imagine a time when the book, as food for the mind and as a vehicle of tradition, hardly existed for the overwhelming majority of the population. In his *Waning of the Middle Ages*, J. Huizinga has painted a vivid picture of what the fifteenth century was like in this respect. Curtius has described by what textbooks the young in the grammar schools were taught throughout the Middle Ages and on what yet more remote sources these textbooks, in turn, drew. But each one of us can remember from his own experience how invaluable for a child's development are the books which he has at his disposal outside the classroom— these are the books which matter most—for the free and joyous satisfaction of the soul's curiosity, at different times and in different classes of society. Goethe has given us, in *Dichtung und Wahrheit*, a wonderful description of his own youthful reading; with obvious delight, which we can understand perfectly if we think back to the bliss of our own first childhood reading, he tells of the earliest food of his mind: the great folio Bible, *Orbis Pictus* by Amos Comenius, Fénelon's *Télémaque*, *Robinson Crusoe* and *Till Eulenspiegel*, *Die schöne Melusine*, *Kaiser Oktavian*, *Fortunatus*, and the whole tribe down to the *Wandering Jew*. In earlier centuries, children were not so lucky because there were no children's books, and apart from the Bible, they had to fall back on Ovid, Statius, or Virgil. It is easy to appreciate how very significant this must have been for the transmission of the European literary tradition.[11]

This tradition, there can be no doubt whatever about it, is seriously threatened today, if indeed it is not drawing to its close. Without embellishing anything and leaving modernistic nonsense to one side, we have to admit that this tradition is collapsing and that its collapse is burying priceless treasures. One of the first and indisputable consequences is perhaps the decay of our languages.[12] But why, we may ask, is the loss or even the dilution of this Christian and humanistic cultural tradition more than a change of scene in the history of thought? Why is it a cultural catastrophe, which is of the essence of our present cultural crisis? Because this tradition is a European tradition and because it makes us Europeans in the widest sense of the word. What this means can easily be appreciated by anyone who merely tries to imagine what the world, a world in which every continent is built upon Europe and its traditions, would be like without this pillar.[13] We cannot even seriously conceive of the idea that after three thousand years we should have to begin again at the beginning in fashioning our minds and that we could possibly replace our spiritual heritage by educational matter of the kind which may roughly be indicated by the range and style of popular magazines, that is, by run-of-the-mill knowledge and run-of-the-mill discussions about vitamins, jet aircraft, social questions, the *dernier cri* of literature, and the latest creations of philosophy. What happens when the attempt is nevertheless made is precisely what the modern world is so eager to demonstrate.

The question that faces us is frighteningly clear. It is the question of whether the fight for existence in which the European cultural tradition is now engaged is not so desperate precisely because it is simultaneously a fight against the most powerful and menacing forces of our social development. This tradition has, in the eyes of our mass epoch, two things against itself: the fact that it is "tradition" and that, necessarily, it is not within everybody's reach, or better, that it presupposes an intellectual hierarchy of people who are able and willing to make a determined effort to acquire it, develop it, and partake in it. It is as much a challenge to the unstable reforming spirit of the *rerum novarum cupidi* as it is to social resentment, which cannot tolerate a minority being in any way privileged in relation to the mass, least of all when the privilege rests on the inexorable exclusivity of personal talent, gifts, and aspirations and is, as such, far more embittering than sheer material wealth, which is not, in principle, out of reach in the contest and lottery of economic life. To all this we must add modern technical and pseudo-scientific pragmatism and utilitarianism and their total inability to grasp that the achievements of the natural sciences, important and formative though they are, cannot change man's nature as primarily a spiritual and moral being. All in all, we can hardly be surprised that the consumptive disease of our cultural tradition has reached the galloping stage in our generation. At the same time, historical awareness, the sense of continuity and of our links with history as a living part of knowledge, is declining more and more widely. This, too, is an essential feature of a faithful overall picture of our modern mass culture.

In conclusion, we want to dispel a common misconception on a point regarding which even John Stuart Mill's famous essay *On Liberty* is dangerously misleading. It is very easy to stress only the conformism and uniformity of our mass culture and to imagine that they imply the complete absence of the courage to be eccentric, unconventional, individual. But the situation is not like that at all. Certainly there is conformism and uniformity, but they are displayed precisely in breaking with the cultural tradition and in trying, at all costs, to proffer something new, original, or provocative *pour épater le bourgeois.* There is not as much conformism in tradition as there is in willful eccentricity; it goes with disorientation and discontinuity, with disdain of anything conventional, time tested, or normal, with sophistication and admiration of the *avant-garde,* with the cult of whatever happens to be new today and discarded tomorrow, the idolizing of novelty. It is conformism in being non-conformist. As an American critic of mass culture puts it, the modern Babbitt thinks he owes it to himself to hang a Picasso on his wall.[14] But "bitter" *Kitsch* is no better than "sweet" and is just as revealing.

63

Indeed, nothing gives mass man away as much as his contempt for what he calls "suburban," "bourgeois," or "Philistine." His conformism manifests itself in the break with tradition and continuity, in rootlessness, anti-conservatism, revolutionary romanticism. This is the main danger today. The opposite danger of stale and stuffy stagnation hardly arises in present circumstances. This view of the situation is confirmed if we look beyond our own immediate horizon to what goes on in the so-called underdeveloped countries. The only hopeful approach is to regard the situation as a violation of human nature and society which is hard to conceive as permanent and which is bound, sooner or later, to end in an acute crisis, which might just possibly have the salutary effect of bringing us back to our senses.

These reflections are not merely the disgruntled and pretentious musings of a reactionary standing in holy terror of anything called mass. Socialists like George Orwell and Hendrik de Man did not think otherwise. In fact, we are indebted to the latter for a particularly grim and penetrating analysis of the relationship between mass and cultural decline.[15]

These men's testimony confirms me in my feeling that this subject is so important that I should not leave it without adding a few more examples to give it color and life. First let me demonstrate how far we have already got with our *tabula rasa* attitudes. I quote a former Director-General of the World Health Organization, Dr. Brock Chisholm. A few years ago, he wrote in a Canadian paper (*Toronto Telegram, Weekend Magazine* [1955, No. 10]) that we should create a uniform world population by mixing all races. "The sooner we're all interbred, the better," says this man, who, grotesquely, was put in charge of world health. He wants to do away with all tradition, as well as with "our economic and social structure," and thereby acquires at least the distinction of pushing a sort of super-mass progressivism to hitherto unheard-of lengths. His example naturally carries special weight because of his office.

My next witness shows that in an advanced mass society there is no longer any feeling for inevitable differentiation in upbringing and manners. An American girl student, typical in this respect, declared her utter incomprehension when a working-class Frenchwoman remarked of some young lady that she was *bien élevée*. Nor did the student understand what I meant by my ironic retort that this must be because all girls in her country were *bien élevées*. Obviously, the concept as such—which is none other than Pascal's *honnête homme*—no longer exists in a mass society. Let me add at once, to cut short any possible European self-righteousness, that we have every reason to ask ourselves seriously whether there are not a good many countries on this side of the Atlantic Ocean, too, which are scarcely distinguishable from America in respect to the intellectual domination of the "common man," who is representative not of the "people," spontaneous, natural, and possessed of a sound native power of judgment, but of "mass."[16] Mass culture is everywhere crowding out the two forms of real culture, those of the elite and of the people.

Mass and Society

The mass phenomena of cultural development which we have just discussed can obviously not be neatly separated from those of social development. We should keep this well in mind as we now turn to a closer examination of the relationship between mass society and mass democracy and the final, poisonous product of its decomposition, totalitarianism. This was expressed with the utmost clarity long ago by that classical writer who, with the exception of Pestalozzi, was the first and most profound of the prophets of gloom to analyze in advance modern mass democracy, namely, Alexis de Tocqueville. I have in mind especially the famous passage in his great work *De la démocratie en Amérique*, where, groping his way on untrodden ground, he tried to show how egalitarian democracy was bound to develop into a new form of despotism. His only mistake, and a pardonable one, was to imagine this new mass-democratic despotism as something comparatively mild, whereas we know better.[17]

It would require a long treatise to describe the elements and conditions which mark out and determine the path leading to mass democracy and its extraordinary dangers to liberty. The essential point becomes clear if we consider the difference between liberal democracy of the Anglo-Saxon and Swiss kind on the one hand and the Jacobin brand of democracy on the other. The latter has increasingly become the dominating form of democracy in our times precisely because it is appropriate to mass society. Why is this so, and what is the link between Jacobin democracy and mass society? If we say that liberal democracy places the accent on liberty and Jacobin democracy on equality, this means in practice that the former rests on government with the consent and under the control of those governed and the latter on the principle of the sovereignty of the people, ascertained by majority decision and intended to realize the identity of people and government.

Now while this Jacobin sovereignty of the people is a fiction, it is a highly dangerous one because it opens the way to the worst despotism and makes it possible for a majority decision to establish a totalitarian government. Liberal democracy is a source of freedom because it is liberal, that is, respectful of the individual's right to liberty, and because it is, at the same time, democracy, that is, makes government subject to the consent of those governed. Jacobin democracy, however, is the ultimate ruin of freedom. Now this Jacobin ideology of the sovereign people is less harmful when the state is regarded as something alien, something to be watched with suspicion. This is the case in federal constitutions, where the remote central government is not locally regarded as one's "own" and therefore stimulates liberal defenses, while at the same time the central government acts as a counterweight to any despotic separatism of the members of the federation at all levels. It is for this reason that the Jacobins always rightly judged federalism to be the worst enemy of the *république une et indivisible*. But since a federal political structure cannot be created simply by the fiat of a constitutional document but presupposes a social articulation rooted in tradition, nothing is more inimical to federalism than modern mass society because it destroys this articulation. Thus mass society fosters Jacobin mass democracy by paralyzing and destroying the countervailing forces of federalism. Even where these still survive they are clearly on the defensive, and their prospects of victory are constantly deteriorating. The outlook is, perhaps, most favorable in Switzerland because that country's multi-national character makes federalism vital to the existence of the state.

In Germany, on the other hand, federalism seems to have no live roots any more, its long tradition notwithstanding. This is one of the most impressive symptoms of the alarming degree to which that country has already become a prey to enmassment after having passed through the mill of totalitarianism and defeat. Even the circles most vociferous in their condemnation of mass society remain blind to this unequivocal symptom, as is demonstrated by their disdain of federalism. Witness the ease with which recently, under their leadership, or at any rate with their consent, Württemberg and Baden were wiped out as historical entities and, like two factories, merged in the name of administrative convenience— though even that has proved elusive.

It is therefore hardly surprising that some years ago a socialist minister in one of the German *Länder* should have declared, in all seriousness, that federalism, autonomy, and local government were quite unnecessary in a democratic state because, in contrast to the old authoritarian state, there was no longer any division between people and government. These things, he said, had been an expression of justified mistrust toward a central government divorced from the people and as such were out of place today. This statement at least serves to destroy any cherished illusion about this radical Jacobin ideology and its consequences. One is reminded of the Communists, who, when reproached with the total absence of independent trade-unions in Soviet Russia, reply ingenuously that in a state where government and working class have merged into one, there is no need for any independent trade-unions to safeguard the workers' interests. This is, in fact, what goes on in the minds of the Jacobins, and they are all under the spell of a myth whose pseudo-religious content is unmistakable—even without the recent incident when an election poster of the Swiss Communist party blasphemously put the "sovereign" people in the place of God and parodied the Bible by declaring: "The fear of the people is the beginning of wisdom." It is ominous that much the same used to be said by the leaders of National Socialist Germany.

The close and indeed indissoluble link between mass society and Jacobin democracy can perhaps be seen even more clearly if we look at it in another light. What distinguishes mass democracy of a Jacobin complexion from liberal democracy is that the former, with its emphasis on the sovereignty of the people, does not acknowledge the decisions of the "sovereign" to be subject to any higher and absolute authority, such as liberal democracy recognizes. For the latter, there are certain limits to the power of the state which may not be transgressed by any popular or parliamentary majority; they are the limits traced by the traditional principles of government, the unchallengeable commands of ethics and natural law, and the unwritten precepts of the history of nations.

Anglo-Saxon and Swiss democracy are rooted in historical soil that is centuries older than universal suffrage; they grew up in an age when the ancient elements of freedom, whether of classical, Christian, or Germanic origin, were still a live reality and when the area of rights and obligations was firmly circumscribed by a society whose fabric and structure were the very opposite of modern mass society. Most of us still possess enough good sense to recognize liberal democracy's essentially metademocratic limits in extreme cases and to take it for granted that we cannot leave it to a majority decision whether our country is to become Communist. But there are few who are aware of the implied perilous fragility of the undiluted democratic principle, which exposes us all along the line to arbitrary power and to the dissolution of the inviolable principles of government and society and which can protect neither freedom nor property nor law from despotism. It is, as Jacob Burckhardt said, the "end of all security."

Democracy is, in the long run, compatible with freedom only on condition that all, or at least most, voters are agreed that certain supreme norms and principles of public life and economic order must remain outside the sphere of democratic decisions. This *unitas in necessariis* encompasses more than the principle of the rule of law, which, though admittedly important, is ultimately only formal. It is this fundamental agreement which imbues the concept of inviolable law as such with an absolute content, and once it can no longer be taken for granted, we are in the presence of mass democracy of a pretotalitarian kind. We hardly need the lessons of the Jacobin government of the French Revolution to remind us forcibly of the inevitable socialist tendencies of such mass democracy. Conversely, socialism is fatefully ranged among the forces responsible for the transformation of liberal democracy into mass democracy. To divest such fundamental institutions as property and economic freedom of their inviolability and to make their fate dependent on the ballot box is tantamount to destroying the very foundations of liberal democracy, since the latter must rest on above-party agreement with respect to the unchallenged validity of the state's ethical, social, and political principles. Democracy is then no longer an instrument of internal peace, security, stability, and freedom but becomes a tool of revolution, and of permanent revolution at that.

This kind of radicalism, typical of a spirit which is not content to accept what is but must forever reopen every question, is precisely the mark of mass society and mass man. It is the spirit of men who, together with their social roots, have lost the sense of tradition, principles, and history and who have become the prey of the moment's whims and passions, as well as of the demagogy of leaders translating these whims and passions into ephemeral slogans

69

and inflammatory speeches. Matters are yet made worse by the skepticism and positivism, sometimes bordering on nihilism, for which the intellectuals are to blame. Thus modern mass democracy becomes the breeding ground for the revolutionary social religions of our times and the rallying point for the crusades on which the inflamed masses set out to conquer some millennium, some New Jerusalem. What Michel Chevalier *(Lettres de l'Amérique du Nord* [1836]), at the same time as Tocqueville, rightly or wrongly said of the American people applies without reservation to mass democracy: it has "the morale of an army on the march." And it has all the attributes of such an army, too: the clockwork precision of the march, collective high spirits, noise, the "here today and gone tomorrow" outlook, the moment's enjoyment without thought of the future, the transience of life, the rousing banners, a nomad and rootless existence, practical action unburdened by disdained theory, pragmatism, grabbing and wasting, and good-fellowship.[18]

We shall have more to say later about the significance for today's economic and social life of this mass democracy cut loose from the moorings of natural law and tradition. Meanwhile, we want to illuminate from yet other angles our main subject of "mass and society."

Let us return once more to Tocqueville. It is to his credit that, unlike his successor, Mill, he clearly recognized that it would be entirely wrong to regard as two opposite poles mass society and the individual, or mass behavior and individualism. On the contrary, mass man *is* individualistic because of the loosening of the social fabric and the disintegration of community. One of the least well understood aspects of the process of enmassment is that it detaches the individual from his natural social fabric and leaves him to his own resources. Conversely, when individualism became the cult of the isolated individual as such and of the mere millionfold voter, in contrast to true community and social articulation, to *corps intermédiaires* and federalism and political and social pluralism, it also became one of the most corrosive of spiritual acids, dissolving the organic structure of society and thereby contributing to the formation of mass society and mass democracy.

Mass society is simply a sand-heap of individuals who are more dependent than ever, less sharply defined and more depersonalized than ever, and at the same time more isolated, uprooted, abandoned, and socially disintegrated than ever. This is what we must grasp if we want to understand the true nature of mass society and its political, spiritual, social, and economic consequences. *The Lonely Crowd*—nothing could describe the phenomenon better than these words, which aptly form the title of a book by the American sociologist David Riesman. It is true that solitude, in the sense of communing silently with oneself and nature, is becoming both rarer and less earnestly desired. But loneliness, separateness, and isolation are becoming the destiny of the masses, and this is a situation which is so highly pathological that one is tempted to agree in this respect with certain serious authors who regard our modern world as afflicted by collective lunacy. This was the diagnosis pronounced by the Dutchman J. Huizinga even before the Second World War, and since then, psychiatrists have repeated the warning. It is true that the isolation of the individual has several concurrent causes, but the most important among them is undoubtedly the mass-conditioned disintegration of community.[19]

It is significant, and perhaps also comforting, that it should be another American, Ernest van den Haag (with Ralph Ross in *The Fabric of Society* [New York, 1957], Chapter 15), who has given us the most disillusioned account of the depersonalization and social disintegration going on in an advanced mass society. People are divested of their natural individuality, both as producers, turning out mass products by more and more mechanized methods, and as consumers, because mass products cannot cater to individual tastes and because, at the same time, the class of people wealthy enough to buy custom-made goods is dwindling fast, thanks to invidious taxation imposed by the crushing majority of consumers with "shared tastes." Impersonal work has its counterpart in

71

impersonal consumption; the standardization of objects demands and eventually enforces the standardization of persons. "Most people perch unsteadily in mass-produced, impermanent dwellings throughout their lives. They are born in hospitals, fed in cafeterias, married in hotels. After terminal care, they die in hospitals, are shelved briefly in funeral homes, and are finally incinerated. On each of these occasions—and how many others?—efficiency and economy are obtained and individuality and continuity stripped off. If one lives and dies discontinuously and promiscuously in anonymous surroundings, it becomes hard to identify with anything, even the self, and uneconomic to be attached to anything— even one's own individuality. The rhythm of individual life loses autonomy, spontaneity, and distinction when it is tied into a stream of traffic and carried along according to the speed of the road, as we are in going to work, or play, or in doing anything." A modern Faust, enjoining the fleeing moment to "linger a while, thou art so fair," would not lose his soul but cause a traffic jam.

How right, too, is this acute American observer in his remark about the frantic and hopeless efforts of the "human-relations industry" to reconstitute individuality artificially—"scientifically" is the word, of course—by pseudo-personalization. These efforts are on a level with the growing practice of the food industry, which first extracts all the goodness from flour and then enriches it with artificial vitamins. Even the name card by which the official at the ticket booth is personally introduced to us merely proves how impersonal our contact is; in the natural conditions of the village or small town such a name card would be entirely redundant. It is life affectedly giving itself scientific airs, like the popular craze for vitamins or the enticing inscription "Ride for your health" which I once saw at the Atlanta airport on a rocking horse placed there to keep children amused. This, of course, was meant as advertising, and we are anticipating a later chapter.

These socially disintegrated individuals are like physical bodies with peripheral centers of gravity. Hence another significant symptom: the spread of a harrowing sense of *eccentricity*. Less and less men are at rest within themselves, or, to quote a famous passage from Jacob Burckhardt's letter to Friedrich von Preen of December 30, 1875, "they shiver with cold unless they are huddled together in their thousands." This is true not only of single persons but of whole groups, of the small and outlying groups, that is, in relation to large and central ones, in particular of the village and small town in relation to the large town and metropolis. In these smaller communities, people are increasingly losing the last vestige of what ethnologists call ethnocentrism, to be found in its pure state among tribes that are thus far untouched by our civilization. On the contrary, more and more people are becoming "ethnofugal," haunted by the feeling of "being out of things" and tormented by the fear of becoming "rustic and provincial," as they significantly call it, unless they form part of the great human heap of the metropolis.

The result is an attitude which one might describe as our era's characteristic *alibism*, that is, an instinctive craving to be elsewhere and a neurotic feeling of being always in the wrong place. This is powerfully fostered and at the same time technically facilitated by the automobile, radio, films, and television, and the resulting nomadism of our day takes multiple forms: the apparently irresistible appeal of large cities; a yearning for gregariousness, even in one's hours of leisure; constantly being on the move; an increasingly pathological kind of mass tourism that finds its be-all and end-all in "doing" sights, taking snapshots, and "having been there"; and "congressitis" and the mania of meetings and festivities.

Nor is this all. The great problem of mass society is not exhausted by isolation, emptiness, and loss of solidarity, together with the ensuing void in the normal human life's conditions of happiness. Something else is almost more oppressive and decisive, something which is far from being given the attention it deserves: this is the *boredom* of mass society. Boredom is a product and an accessory of mass society and at the same time the cause of things which in-

73

tensify some of its features and drive it still further. No thoughtful person can deny that here is one of the master keys to an understanding of our modern world. Boredom is the true curse of our epoch—boredom, together with mass living, the isolation of the individual and his separateness, the weakening of society's inner mainspring, and the triumph of vapid utility over the poetry, dignity, spontaneity, and grace of life.

Boredom and Mass Society

The vast and important subject we have now reached is so contested that we must go a bit further afield, for what we now have to say touches upon the very foundations of the irrepressible optimists' position. Again and again it becomes evident in discussions of mass society and the cultural crisis that what we regard as the central issue of our epoch and the cause of its being one of the most critical periods of world history is not seen by others in the same light. For us, mass society and the giant strides with which technology advances through our world are the symptoms and sources of a severe disease of society and of a fatal alteration of the individual's spiritual and moral conditions of life; in our view it is here that we must seek the essential causes of the threat to liberty and personality in the shape of collectivism and totalitarianism. But we have learned to expect that someone will always come forward to deny this diagnosis in the name of progress and liberty. Indeed, some declare the symptoms of a truly vast disease to be those of recovery, holding out equally vast promises.

It would be quite wrong simply to consider such views as "American," as contrasted with "European" views. It is true that the American pattern of life and thought, largely because it is proper to a society less able to draw on the reserves of a healthier past, in many respects displays most clearly some of the things which cause misgivings to the critic of mass society and mechanization. Any American who is determined to remain an apologist to the last is thereby misled into optimism and social rationalism. But we need only recall names like George Santayana, Russell Kirk, or Walter Lippmann and the army of their spiritual kin, the neo-conservatives or decentralists. They demonstrate how, under the impact of America's extreme experiences, the questionable and indeed alarming aspects of this pattern of life and thought call forth the criticism of the best Americans and mobilize moral and religious reserves such as one would be only too glad to see among some Europeans who persist in their self-satisfaction or blind intent to imitate America. On the other hand, there is no lack of Europeans trying to outdo the transatlantic apologists of Americanism and of what they call "American freedom" and who, with an optimism hard to understand, turn a deaf ear to our criticism of mass society, mechanization, and technolatry.

Fortunately, then, there is no conflict between what might be described as European and American mentality. Our European super-Americans, with their eulogies of the magnificent "consumer society" or whatever else they may call it, only cover themselves with ridicule, not least in the eyes of the Americans, who know better and who regard European enthusiasm for "Americanism" merely as disloyalty to a common patrimony. The conflict is, instead, between two social philosophies, tied neither to nations nor continents nor social classes but reaching deep down into the substrata of religion. Once more we are reminded of Cardinal Manning's words: "All human differences are ultimately religious ones." In any event, the conflict clearly does not lack a pronounced political flavor, which comes out clearly when the critique of mass society is dismissed as reactionary—one argument among many common to progressives and Communists.

At first sight it seems difficult to discover any bridge of mutual understanding or even a ground for fruitful discussion. To brand people of our kind as romantic (a designation which may soon become an honorable one) is no more helpful than for us, in our turn, to dismiss our opponents as anthropologically or sociologically blind optimists or social rationalists.

How can we break out of this sterile position of mutual non-understanding? Is it possible to find a sound basis for objectively acceptable arguments against which the optimists can raise no protest?

It is of the utmost importance to realize that such a possibility does exist. Let us try to make a start. There is a certain and indeed predominant type of optimist who behaves like the blind talking of color—and often rather aggressively. Being himself the creature of a large industrial city, he lacks the experience of people born and bred in the country, with an intimate knowledge of village community life and closeness to nature—the very opposite of what Hans Freyer calls "man-made society." There is an asymmetry between people of the former kind and those of us who have had the good fortune of such experience; they are poorer than we are in the knowledge of the things which are at issue in this discussion and provide us with a true yardstick. We who come from the country—or who have at least remained in close contact with country life and have become town dwellers only in our later years—have the advantage of knowing both environments. The others are familiar only with the urban and industrial one. A second asymmetry is this: hardly any country-bred inhabitant of a big city has accepted the urban and industrial reforming optimism and the rootlessness of mass society as an ideal, whereas among those born in the city and the landscape of industry the number is legion who are under no illusion about the price they have to pay in the shape of loss of integration with community and nature.

It may well be worth pondering this state of affairs a little in order to discover which side is the more stricken with blindness and the poorer in experience. It follows that in any discussion we are entitled to ask where our optimist adversary comes from and whether he possesses the same breadth of experience as we do; if he cannot adduce proof to that effect, he ought to admit that he is not competent to speak.

Unfortunately, however, we here encounter a further circumstance which can but reinforce the pessimism of the critics of industrial mass society. The more urbanized industrial man becomes the predominant type, the greater is the likelihood that urbanization, mass living, and the advancing mechanization of life and landscape will be treated as items of progress and liberty by those intellectuals who, born in the same environment, have nothing but irony and arrogance for those who really know better.

We therefore have to envisage the depressing prospect that a protesting voice may find it increasingly hard to make itself heard at all—not because our society is doing so gratifyingly well but because its disease may go so far that we lose any sense of what is or is not health, which would not leave people feeling any happier or healthier.

In all of these respects a country like Switzerland is still comparatively well off because the percentage of people who have grown up in the country or who are at least connected with it by something more than tourism is still unusually high. "It is no accident," the Bavarian critic Joseph Hofmiller writes ("Form ist Alles," *Aphorismen zu Literatur und Kunst* [Munich, 1955]), "that the Swiss have such beautiful children's stories: they do not inhabit large towns. A metropolitan child doesn't even know what it means to be a child. To be a child means to play in the fields, amidst grass and trees and birds and butterflies, under the endless canopy of a blue sky, in a great silence in which the crowing of the neighbor's cock is an event, as is the Angelus bell or the creaking of a wheel. To be a child means to live with the seasons, the first snow and the first colt's foot, the cherry blossom and the cherry harvest, the scent of flowering crops and dry grass, the tickling of the stubble on one's bare feet, the early lighting of the lamp. The other thing is a surrogate, shabby, cramped, musty, an adult's life *en miniature*." In a cruel cartoon of a Berlin back yard by Heinrich Zille, a porter chases children away from a miserable potted plant and calls after them: "Go and play with the dustbins!"

Without people feeling any happier or healthier, I said a little earlier, and these words bring me to the essential point. It can be

77

described quite simply, in the words of the Gospel: "For what is a man profited, if he shall gain the whole world, and lose his own soul?" Or less solemnly: our critical attitude to mass society and technolatry is superior to the optimistic attitude because the latter undeniably leads to consequences of an unequivocally negative kind. Uncritical optimism turns out to be inadequate anthropology because it fails to consider how man will fare, both in his body and, above all, in his soul. The uncritical do not grasp what ails people in the much-vaunted conditions of today, in spite of tiled bathrooms, macadamized roads, and television. Even the optimists cannot overlook the fact that dissatisfaction and discontent seem only to grow with the profusion of goods designed for creature comforts and in inverse proportion to the happiness expected of those goods. But they are unable to understand the profound causes of this apparent paradox.

Our real condition can best be judged by those who have the most direct access to the body and the soul of man and who penetrate his façade. Foremost among them are the ministers of the church, but their testimony is not public and lacks the compelling force of persuasion which might disarm doubters. But the testimony of physicians—on both sides of the Atlantic Ocean—is unanswerable. When the consulting rooms of psychiatrists, neurologists, and heart specialists fill up with the wreckage of our civilization, no paeans extolling motorcars and concrete will help. "Only myopic apostles of progress," writes a German psychiatrist, "can still deny that our technical world of artifice threatens to become a deadly menace to man. His spiritual and bodily constitution is being drastically modified by this world, to which man has abdicated essential and inalienable elements of his nature in order to keep the machines running. The functioning of the mechanism has become autonomous; man has personified technology and its paraphernalia and has thereby depersonalized himself."[20] The objective language of the statistics of heart and nervous diseases, of suicides and the consumption of tranquilizers, should be clearest precisely to the people who are quantity-conscious, and if it is not heeded earlier, it certainly will be when we are all caught up in these macabre classifications. This language must surely be intelligible to those who either do not hear or cannot interpret the more subtle language of artists reacting to their environment with enhanced sensibility.

Thus there is no lack of warning signs which should call us to our senses. A little honest introspection will force us to admit that mass society and industrial and urban civilization are threatening to condemn us to conditions of life which are simply beyond the human scale. No amount of modernism is of any avail against this stark fact, no social eudaemonism, and no anathema against "reactionaries" and "romanticists."

When people today react to their environment by feeling vaguely discontented or even unhappy, the explanation is sometimes sought in fear and anxiety. The philosophers of existentialism have built an entire system on *Angst*. No one will deny that fear and anxiety are deeply lodged in our world today; they are those evil spirits of which the Gray Woman, Care, spoke to Faust:

Whomsoever I possess,

> Finds the world but nothingness;
> Gloom descends on him for ever,
> Seeing sunrise, sunset, never;
> Though his senses are not wrong,
> Darknesses within him throng,
> Who—of all that he may own—
> Never owns himself alone.
> Luck, ill luck, become but fancy;
> Starving in the midst of plenty,
> Be it rapture, be it sorrow,
> He postpones it till to-morrow,
> Fixed upon futurity,
> Can never really come to be.[21]

Fear and anxiety can destroy man only when the meaning and purpose of his life have become blurred or escape him. In the words of the British writer Charles Morgan, one of the few lucid and noble minds of our times: "Neither suffering nor even terror produces despair in men; loneliness and boredom are the prime cause of it, and they are the miseries that beset our crowded and eventful age."[22]

One of these miseries, loneliness, has already been discussed sufficiently here and elsewhere. All the more need is there to turn our full attention to the other, boredom. Much has been said about boredom as a universal and perennial affliction of mankind, but little in the particular context of its being the product and curse of mass society. What Pascal and Schopenhauer said about it is still eminently worth reading and instructive, but our own era has taught us that this spiritual disease, which is closely akin to the accidie of the medieval Church, may have its essential origin not in the aberrations of the individual soul but in the conditions and influences of society. Boredom is what Georges Bernanos' country parson, at the beginning of *Journal d'un curé de campagne*, describes as "the fine dust which today settles on all things, not sparing even the countryside, and against which men try to defend themselves with their excited bustle."[23]

We could find no better motto for our considerations here than the following passage by a contemporary ethnologist writing about a tribe on one of the Pacific islands: "The natives of that unfortunate archipelago are dying out principally for the reason that the civilization forced upon them has deprived them of all interest in life. They are dying of pure boredom. When every theater has been replaced by one hundred cinemas, when every musical instrument has been replaced by one hundred gramophones, when every horse has been replaced by one hundred cheap motorcars, when electrical ingenuity has made it possible for every child to hear its bedtime stories from a loudspeaker, when applied science has done everything possible with the materials on this earth to make life as interesting as possible, it will not be surprising if the population of the entire civilized world follows the fate of the Melanesians."[24]

Even if not taken literally, this outburst points the way to the source of boredom as a social phenomenon. There are many causes, all extremely subtle individually and all so closely interwoven that they cannot easily be disentangled. With these reservations, I would venture the following analysis.

The *first* point is this. It seems reasonable to attribute the mass reaction of boredom, often enough carefully concealed, to a number of features peculiar to modern mass society, such as the loss of communal interests, the disappearance of diversity and spontaneity, emptiness, and isolation.

The dissolution of the natural social order, the inner emptiness of mechanized and quantified work, and the general loosening of the roots of life drive people all the more to fill their time with so-called pleasures and amusements. But they soon discover that they are merely exchanging one kind of emptiness for another because they have lost the meaning and purpose of life. The same civilization whose modern production techniques shower people with the means of comfort and entertainment robs them, at the same time, of any personal relation to their own work. And if people thus cheated of a genuine interest in life seek compensation in consumption, they are fooled once more. The naïve calculation of the sociologists—who persuade us not to grieve over the changed nature of work and to look for compensation in the blessings of "leisure" and "consumption"—this calculation never works out and the sociologists do not seem to understand why.

Let us listen to a frank description of life in a German industrial town: "A small group of sixty thousand industrial workers is here situated in the midst of delightful birchwood country and turns its back on nature—and not only on nature, but obviously also on reason, customs, and experience. . . . There is nothing to give resilience to this community of haphazard newcomers, settled in new housing developments; it has forgotten not only that we lost the war and lost our industry, it seems to have forgotten also that human life needs a focus. In the highly civilized and specialized labor world

81

of Marl, 100 per cent of the men and 20 per cent of the women who are fit to work meet every day in a performance whose ultimate meaning is understood only by very few. . . . The loss of the meaning, of the visible and tangible meaning, of work becomes obvious in such industrial towns. There is no peasant who cannot tell the yield of each field, of each head of cattle, of each tree. But the worker in this super-mechanized world cannot do so any longer. Somehow he has to be helped over the loss of life's visible meaning, and it must be admitted that every effort is being made here to this end. At Marl, free time is a conglomeration of disjointed sensual and intellectual attractions. There is the non-political local newspaper, radio, movies, and television. There are magazines, sports news, and illustrated papers. There are endless books from lending libraries or reading clubs. There are technical gadgets, the motorized bicycle and the motorcycle, the whirring pinball machines . . . but nowhere true leisure, never true contemplation. What, indeed, is there to contemplate? Nature around Marl has become stage scenery; it isn't one's own any more. Nature can be seen in color almost as natural in the technicolor films. So people search for the nature they have lost by going further and further afield on their holiday trips."[25] The same author tells us of an old Ruhr workman who said to him: "In the old days every miner had the ambition to become a foreman; today all he wants is higher wages and shorter working hours." Thus it becomes possible to make such jokes as that of an American cartoonist who depicts a Congressional candidate canvassing for votes with the declaration that although his opponent had promised a four-day week, he had given no assurance that these long hours of leisure would be agreeably occupied.

Now we come to the *second*, closely connected point: the stultifying effect on life of utilitarianism, economism, and materialism, on which we shall have more to say later. "A society which concentrates on material gains will be at once immensely productive and immensely sterile, satiated and hungry, busy and enormously bored."[26] Long ago, Tocqueville, in his wisdom, recognized that here was one of the great dangers of mass society, a danger which might easily draw us into an inescapable whirlpool. "Democracy encourages a taste for physical gratification; this taste, if it becomes excessive, soon disposes men to believe that all is matter only; and materialism, in its turn, hurries them on with mad impatience to these same delights; such is the fatal circle within which democratic nations are driven round."[27] But the real cause of this vicious circle had escaped Tocqueville; it is the boredom of a society devoted to physical gratification and driven by its boredom to ever further and, above all, ever new enjoyments. "It is the boredom of a spoiled child who has too many toys, who can get everything without effort, whose every wish is fulfilled. Television is a new toy of some attraction for all those bored people who already own a car and a radio and oil heating and a fully electric kitchen and who knows what else besides; but it will relieve their boredom for only a short while and afterwards intensify it all the more."[28]

All of this is serious enough, but in actual fact, the evil is seated even more deeply and reaches a level of which the spokesmen of modernism are probably not even aware. The question which we have to ask ourselves is this: Is it not a fact that day after day and with immense energy and equally immense infatuation we are busy creating a material environment which suffocates the soul of man and causes psychical lesions of an immeasurable and incurable kind? And is it not a fact that we do this in the name of bare material utility and in the service of measurable economic gain, without even noticing that we are causing enormous damage to higher things, damage which may well have a decisive effect on our own lives? There is a downright uncanny power in our modern industrial, urban, and mass civilization which destroys all beauty, dignity, harmony, and poetry in its path, so much so that it has justly been called *The Ugly Civilization*, to quote the title of a book by R. Borsodi. The modern world of concrete, gasoline, and advertising is peculiarly apt to deprive our souls of certain indispensable vitamins—Burke's unbought graces of life again—and it does so in the name of a technological and social rationalism which has no use for anything that just happens by itself or that is not planned, that grows wild in picturesque confusion, and whose effects defy measurement.

We violate nature at every turn, even to the total disappearance of the countryside, which was recently hailed by a German physicist as the dawn of a new era. We have already seen that we do this at our own peril as far as biological reasons are concerned. Now we have to acknowledge that we are at the same time interfering with the soul of man and depriving him of an essential vital force.[29] Inescapably, we lack an essential spice of life and feel that everything is unaccountably insipid, if we meet only people everywhere and human artifices instead of nature, if we have no regard for tree or beast and treat them like materials or machines, and if we rob nature of her mysteries until we take pride in even making the weather—by majority decision or otherwise. Africa's magnificent wild life is degraded into a mass target for wealthy amateur sportsmen, and the day may not be far off when we shall be able to show animals to our grandsons only in picture books or in the zoo. One species of birds after another, except the very commonest, capitulates before man; the rivers, streams, and marshes, in so far as the sewage of industry and mass society does not turn them into stinking cesspools, are made into drainage canals, and one valley after another is submerged under the reservoirs of power stations so that more men can shave with electric razors or kill time in front of television screens. Who will dare to maintain that all this can fail to make the world unbearably dull?

Tabula rasa, the domination of the drawing board and sovereign contempt for everything that has grown—this is how we treat not only the landscape of nature but also the cultural landscape of cities. Disregard of nature is here matched by disregard of historical beauty and harmony. Once more we quote Jules Romains, from his book *Le problème numéro un:* "It may happen that beyond a certain point the destruction of a civilization's physiognomy becomes an immeasurable disaster, an unnoticed loss of vital purpose and vital energy." Who knows what irreparable lesions the destruction of the German cities has caused the human soul and how much it has contributed to the striking advance of mass culture in Germany?[30] Every effort to rebuild with due regard to the links with the past deserves the greatest credit. But it is significant that such efforts have to contend with strong resistance on the part of modernists and are therefore only partially successful. It is, in effect, the destructive spirit of modernism which prevails everywhere and which irreverently disfigures our venerable European cities, with the result that they are becoming just as dreary as American ones.

One of the protagonists of modernism, the architect Le Corbusier, has declared with brutal frankness: "The core of our old cities, with their domes and cathedrals, must be broken up and skyscrapers put in their place." But this is only an extreme formulation of the revolutionary destructive spirit proper to modernism. How powerful this is, is well demonstrated by the admiration in which a man like Le Corbusier is held by all the world. This revolutionary spirit of new beginnings, of *tabula rasa,* and of the blotting out of history, with its naïve enthusiasm for the enlightenment which we have at last brought to the world and which has come to stay—this spirit reminds us of the effusions of fashion magazines showering contempt on the last season's models, without a thought of the same fate awaiting the latest fashion. But on a more serious level, it is obvious that this spirit corresponds to the spirit of mass democracy. *"Il faut recommencer à zéro,"* says Le Corbusier, and thereby he translates into the language of architecture Thomas Paine's dictum: "We have it in our power to begin the world over again."

How long can the countryside and the core of our cities withstand this mass onslaught of concrete and the heralds of "dynamic functionalism"? How strong is the resistance against the idea of a synthetic drawing-board town, an idea hatched even in Switzerland

85

and indeed worthy of our times? How long will it be possible to protect the German *Autobahnen* against the pressure groups of advertisers? The country of the traveler's dream, *Italia Diis sacra,* has heroically lent itself as a guinea pig for a demonstration that with a little effort even the most beautiful country in the world can be made unspeakably ugly anywhere within reach of advertising and the filthy vulgarity of suburbia. And what has happened to the country which until the beginning of the nineteenth century was the arbiter of Europe in refinement of taste and which in the Middle Ages was the fountainhead of Western culture—what has happened to France? It has become a garish fairground of vulgar suburbs and provincial towns.

This development, with its contempt of nature and history alike, leads to an impoverishment of the soul because it acts upon us through all the doors of perception. Not only the visual image assails us, but so does its acoustic echo: the noise which rises from modern mass society and grows to real torture in the din of jet aircraft and helicopters. It is not that mere absence of sound is the ideal. There is also a silence of nothingness, the hush of death, stillness in places where we miss the singing of the farm maids, the village band on the green, the warbling of the nightingale, the fanfares from the church tower, the sound of the post horn, the accordion at dusk under the lime tree, the thump of the threshing flail, the crow of the cock. In the realm of sound, too, there is a "natural" order appropriate to man, and what is bad and a cause of boredom is that we hear road drills and motor scooters but not people singing for sheer joy of life. What is so infernal is the "technical" noise of our times, which ends up by making it a blessing to be hard of hearing. Occasionally we get a chance to listen to something pleasant—folk songs, perhaps, or some attraction of this kind—intended, of course, to promote the tourist traffic; but there again we are cheated of the true savor, and that this is a tribute of commercialism to the "unbought graces of life" is scant consolation.

Thus boredom spreads like mildew over all the places left vacant by the disappearance of those graces, dignities, and things which were genuine, natural, which warmed the heart of man. One of them is love, which the sexual obsession of our times is stripping of its tenderness and poetry and which, like so much else, ends up on the great cinder heap of boredom. If adultery becomes an everyday banality, what need have we of *Madame Bovary* and *Effi Briest*? Among the dying embers is the numinous; in the shape of popular faiths, it still has power to warm. Among them also is genuine popular culture, with its customs rooted in the changing seasons, with the traditional festivals, and much else, of which only Christmas has preserved a last glow for most of us—and even that is more and more outshone by the neon lights of advertising.

One could mention much else besides, but that would take us too far. One final observation imposes itself: the disappearance of so many things, great and small, which lend charm, dignity, and poetry to life deprives writers and artists of rewarding and stimulating subjects and so explains the undeniable impoverishment of modern art, in literature and painting alike.[31]

It need not be said how utterly mistaken we would be to blame the market economy for all these causes of boredom as a social phenomenon. On the contrary, the market economy, with its variety, its stress on individual action and responsibility, and its elementary freedoms, is still the source of powerful forces counteracting the boredom of mass society and industrial life, which are common to both capitalism and socialism. Only, the market economy must be kept within the limits which we shall presently discuss. But a socialist mass society is doomed to irresistible boredom. The principle of organizing and centralizing everything blunts all the instincts of independence and responsibility and thus pushes boredom to its utmost limits, unless one is prepared to grant that in the extreme case of Communism, obsession, fear, hatred, and the hope of deliverance or escape from this desert add a rare spice to life.

As regards the socialist welfare state, it is well epitomized in a recent remark by Bernard Berenson, the American Nestor of modern art historians, on the occasion of his ninetieth birthday: "I do not fear the atomic bomb. If there is a threat to our civilization, it is more likely to come from boredom that will result from a totalitarian welfare state and from the exclusion of individual enterprise and the spirit of adventure."[32] Evidence that he is right is accumulating, and indeed, it would be astonishing if it were not so. What else can we expect but that the Swedes, once famous for their happiness, the people of Gösta Berling, today, in their welfare-state paradise, distinguish themselves by unusually high suicide figures and by other symptoms of a terrifying degree of listlessness, discontent, and boredom or that the number of British people sick of being cheated of the enjoyment of the fruits of their own efforts and ready to escape the dullness of the welfare state by emigration has reached alarming heights?

One final word to those whose retort to everything we have said in this chapter is the reproach of romanticism.[33] It certainly is romantic, if by that term we understand resistance to the destruction of dignity and poetry and the "unbought graces of life." If this is romanticism, we profess it unreservedly and proudly, and we will not allow ourselves to be intimidated or abashed by these would-be masterminds. We do not want to set the clock back; we want to set it right.

There remain some perfectly simple and elementary facts which are unanswerable. Can it be denied that, as Walter Lippmann once said, one can hanker after the rose-clad cottage of one's youth but not after a neon-lighted service station? Can it be disputed that a businessman who wants to sell Christmas cards having some sentimental, if elementary, appeal prints dreamy villages with gaily decorated horse-drawn sleighs or snowy landscapes and not automobiles or garages or a town of concrete blocks glittering with advertising? Is it imaginable that Segantini's well-known painting *Plowing in Oberhalbstein* in the Munich collection should have a tractor in the foreground instead of horses? Would it not be better to ponder such things, and others, instead of dismissing them with a supercilious smile?

There can be no doubt about it whatever: what bedevils mankind in our times, though people do not always fully realize it, is the boredom of a disenchanted and depleted world bereft of its mainsprings. It is boredom, perhaps even more than anxiety, and one may well ask whether anxiety and the philosophy of *Angst* are not themselves the product of boredom. Behind the façade of the modern world stands not only the specter of anxiety, which we have mentioned already, but also one of the other "Gray Women" who spared Faust: boredom. Once this is recognized, the whole philosophy of modernism and progressivism crumbles like rotten tinder. Is it not boredom which drives us hither and thither, like unquiet spirits, and makes us clutch at anything which will fill up the gaping void of our existence?

Once more we return to Burke and his oft-quoted unbought graces of life. The expression occurs in a famous passage of his *Reflections on the Revolution in France,* where we also find this sentence: "But the age of chivalry is gone. That of sophisters, œconomists, and calculators has succeeded." Shall we not prove to Burke that he has done the "œconomists" an injustice? Shall we not dissociate ourselves from the sophisters and calculators? Of what avail is any amount of well-being if, at the same time, we steadily render the world more vulgar, uglier, noisier, and drearier and if men lose the moral and spiritual foundations of their existence? Man simply does not live by radio, automobiles, and refrigerators alone, but by the whole unpurchasable world beyond the market and turnover figures, the world of dignity, beauty, poetry, grace, chivalry, love, and friendship, the world of community, variety of life, freedom, and fullness of personality. Circumstances which debar man from such a life or make it difficult for him stand irrevocably convicted, for they destroy the essence of his nature.

CHAPTER III

The Conditions and Limits
of the Market

The questionable things of this world come to grief on their nature, the good ones on their own excesses. Conservative respect for the past and its preservation are indispensable conditions of a sound society, but to cling exclusively to tradition, history, and established customs is an exaggeration leading to intolerable rigidity. The liberal predilection for movement and progress is an equally indispensable counterweight, but if it sets no limits and recognizes nothing as lasting and worth preserving, it ends in disintegration and destruction. The rights òf the community are no less imperative than those of the individual, but exaggeration of the rights of the community in the form of collectivism is just as dangerous as exaggerated individualism and its extreme form, anarchism. Ownership ends up in plutocracy, authority in bondage and despotism, democracy in arbitrariness and demagogy. Whatever political tendencies or currents we choose as examples, it will be found that they always sow the seed of their own destruction when they lose their sense of proportion and overstep their limits. In this field, suicide is the normal cause of death.

The market economy is no exception to the rule. Indeed, its advocates, in so far as they are at all intellectually fastidious, have always recognized that the sphere of the market, of competition, of the system where supply and demand move prices and thereby govern production, may be regarded and defended only as part of a wider general order encompassing ethics, law, the natural conditions of life and happiness, the state, politics, and power. Society as a whole cannot be ruled by the laws of supply and demand, and the state is more than a sort of business company, as has been the conviction of the best conservative opinion since the time of Burke. Individuals who compete on the market and there pursue their own advantage stand all the more in need of the social and moral bonds of community, without which competition degenerates most grievously. As we have said before, the market economy is not everything. It must find its place in a higher order of things which is not ruled by supply and demand, free prices, and competition. It must be firmly contained within an all-embracing order of society in which the imperfections and harshness of economic freedom are corrected by law and in which man is not denied conditions of life appropriate to his nature. Man can wholly fulfill his nature only by freely becoming part of a community and having a sense of solidarity with it. Otherwise he leads a miserable existence and he knows it.

Social Rationalism

The truth is that a society may have a market economy and, at one and the same time, perilously unsound foundations and conditions, for which the market economy is not responsible but which its advocates have every reason to improve or wish to see improved so that the market economy will remain politically and socially feasible in the long run. There is no other way of fulfilling our wish to possess both a market economy and a sound society and a nation where people are, for the most part, happy.

Economists have their typical *déformation professionelle,* their own occupational disease of the mind. Each of us speaks from personal experience when he admits that he does not find it easy to look beyond the circumscribed field of his own discipline and to acknowledge humbly that the sphere of the market, which it is his profession to explore, neither exhausts nor determines society as a whole. The market is only one section of society. It is a very important section, it is true, but still one whose existence is justifiable and possible only because it is part of a larger whole which concerns not economics but philosophy, history, and theology. We may be forgiven for misquoting Lichtenberg and saying: To know economics only is to know not even that. Man, in the words of the Gospel, does not live by bread alone. Let us beware of that caricature of an economist who, watching people cheerfully disporting themselves in their suburban allotments, thinks he has said everything there is to say when he observes that this is not a rational way of producing vegetables—forgetting that it may be an eminently rational way of producing happiness, which alone matters in the last resort. Adam Smith, whose fame rests not only on his *Wealth of Nations* but also on his *Theory of Moral Sentiments*, would have known better.

All of this has always been clear to us, and this is why we have never felt quite comfortable in the company of "liberals," even when styled "neo-liberals." But for everything there is season. We have been through years of untold misery and disorders which so many Western countries, including, in particular, Germany, brought upon themselves by their disregard of the most elementary principles of economic order. During these years there was a compelling need to put the accent on the "bread" of which the Gospel speaks and on the re-establishment of an economic order based on the market economy. To do this was imperative. Today, when the market economy has been revived up to a point and when even its partial re-establishment more than fulfills our expectations, it is equally imperative to think of the other and higher things here under discussion. That the hour is ripe for this is appreciated by all who are wise enough to sense the danger of stopping short at "bread." It is a sign of the times that those who experience and voice these misgivings have become surprisingly numerous everywhere. They include a growing number of economists in several countries who, independently of each other, are stepping out of the ivory tower of their science to explore the open country "beyond supply and demand."[1] As far as this author is concerned, he is doing no more than returning to scientific work of a kind which he has considered paramount ever since he wrote his book on *The Social Crisis of Our Time*.

To the economist, the market economy, as seen from the restricting viewpoint of his own discipline, appears to be no more than one particular type of economic order, a kind of "economic technique" opposed to the socialist one. It is significant of this approach that the very name of the structural principle of this economic order has been borrowed from the language of technology: we speak of the "price mechanism." We move in a world of prices, markets, competition, wage rates, rates of interest, exchange rates, and other economic magnitudes. All of this is perfectly legitimate and fruitful as long as we keep in mind that we have narrowed our angle of vision and do not forget that the market economy is the economic order proper to a definite social structure and to a definite spiritual and moral setting. If we were to neglect the market economy's characteristic of being merely a part of a spiritual and social total order, we would become guilty of an aberration which may be described as social rationalism.

Social rationalism misleads us into imagining that the market economy is no more than an "economic technique" that is applicable in any kind of society and in any kind of spiritual and social climate. Thus the undeniable success of the revival of the market economy in many countries gave quite a few socialists the idea that the price mechanism was a device which an otherwise socialist economy could well use to its own benefit. In this concept of a "socialist market economy," which Tito seems to want to translate into practice, the market economy is thought of as part of a social system that is best described as an enormous apparatus of administration. In this sense, even the Communist economic system of Soviet Russia has always had a "market sector," although it is undoubtedly no more than a technical device and contrivance and not a living organism. How could a genuine market, an area of freedom, spontaneity, and unregimented order, thrive in a social system which is the exact opposite in all respects?

The same social rationalism is evident in the attitude of certain contemporary economists who, while not open partisans of socialism and sometimes speaking in the name of the market economy, work out the most elaborate projects for regulating the movements of the circular flow of the economy. They seem to be prepared to transform the economy into an enormous pumping engine with all sorts of ducts and valves and thermostats, and they not only seem confident that it will function according to the instructions for use but they also seem to be unaware of the question of whether such a machine is compatible with the atmosphere of the market, to which freedom is essential.

All of these protagonists of social rationalism—socialists and circular-flow technicians alike—have a common tendency to become so bemused by aggregate money and income flows that they overlook the fundamental significance of ownership. The market economy rests not on one pillar but on two. It presupposes not only the principle of free prices and competition but also the institution of private ownership, in the true sense of legally safeguarded freedom to dispose of one's own property, including freedom of testation.

To grasp the full significance of ownership to a free society, we must understand that ownership has a dual function. Ownership means, as in civil law, the delimitation of the individual sphere of decision and responsibility against that of other individuals. But ownership also means protection of the individual sphere from political power. It traces limits on the horizontal plane, and also vertically, and only this dual function can fully explain the significance of ownership as an indispensable condition of liberty. All earlier generations of social philosophers agreed on this point.

But ownership is not only a condition of the market economy, it

94

is of the essence. This becomes evident from the following considerations. We start out from competition. We all realize its central importance for a free economy, but the concept is obscured by a confusing ambiguity. Communist governments, too, claim that they are using competition extensively and successfully. We have no reason to doubt that in the factories of Soviet Russia, the managers, and even the workers and employees, have ample opportunity for competitive performance. And in Yugoslavia, Tito made a whole system of the "decentralization" of public enterprises whereby the latter were divided up into independent and mutually competing units; he seems to regard this system, with some pride, as a sort of "socialist market economy." There can be no doubt that such an introduction of competition into a collectivist economic system may raise productivity. Is this not the same virtue which we have in mind when we ascribe the rapid recovery of the German economy chiefly to the re-establishment of competition?

There is obviously some confusion here, which calls for clarification. The confusion is due to neglect of the dual nature of competition and to the lumping together of things which should be kept strictly separate. Competition may have two meanings: it may be an institution for stimulating effort, or it may be a device for regulating and ordering the economic process. In the market economy, competition is both, and it constitutes, therefore, an unrivaled solution of the two cardinal problems of any economic system: the problem of continual inducement to maximum performance and the problem of a continual harmonious ordering and guidance of the economic process. The role of competition in the market economy is to be mainspring and regulator at one and the same time, and it is this dual function which is the secret of the competitive market economy and its inimitable performance.

If we now return to the question of whether a collectivist economic order can take advantage of competition and thus appropriate the secret of the market economy's success without impairing the collectivist nature of the economic order, we know that the answer depends upon which aspect of competition is meant. Competition as a stimulant is simply a psychological technique that is as applicable in a collectivist economy as in a market economy or, indeed, in any group, be it a school or a regiment or any other. We may even note that as far as the effects of competition on human destinies are concerned, it may, in collectivist systems, be hardened in a way that is unknown and impossible in the market economy. But the other function of competition, which is at least equally important for its economic effectiveness, the function of selection in the area of material means of production, meets with the greatest obstacles in collectivist systems. In relation to people, the carrot and the stick are ruthlessly applied, but it is quite another question whether in collectivist systems competition can accomplish so uncompromising, undeviating, and continual a selection of products and firms as takes place in the market economy.

Even on the unwarrantedly charitable assumption that collectivist public authorities resist the temptation to hush up investment errors and have the honest intention to carry out such a continual selection in accordance with the dictates of competition, they would lack the indispensable criterion. This brings us to the other function of competition: to serve as an instrument of the economic order as a whole and as a regulator of the economic process. Unlike the market economy, the collectivist economy is necessarily debarred from such use of competition because no collectivist system can create the necessary precondition without losing its own identity. This precondition is genuine economic independence of firms. Only on this condition is the formation of genuine scarcity prices for capital and consumer goods conceivable, but there can be no independence of firms without private ownership and related freedom of action.

Thus everything is interlocked: competition as a regulator of the economy presupposes free market prices; free market prices are impossible without genuine independence of economic units, and their independence stands and falls by private ownership and freedom of decision, unimpaired and undisturbed by government planning. No collectivist economy can possibly satisfy the last of these conditions without ceasing to be collectivist, and therefore it cannot enjoy the advantages of the regulatory and guiding functions of competition. To try to arrange such competition artificially would be as absurd for a collectivist system as it would be for me to want to play bridge with myself. It follows that "socialist competition" can, at best, stimulate (economically not necessarily rational) performance but cannot rule and guide the economic process. It is only half of what competition is in the market economy, and we may well ask whether this bisection does not reduce the effectiveness of even that half of competition which alone is accessible to collectivism, namely, the stimulating effect. Be that as it may, it remains a serious weakness in any collectivist economy that competition can, at best, fulfill only one of its functions, and even that less than optimally. And it is the incomparable strength of the market economy that it alone can take advantage of the dual nature of competition, which is genuine and fully effective only when it is whole. Just as unavoidable limitation to one aspect of competition gravely handicaps collectivism, so does the combination of both aspects of competition give the market economy a start which cannot be overtaken. This is the prerogative of the market economy, but this prerogative stands and falls by private ownership of the means of production.

The economic function of private ownership tends to be obstinately underestimated, and even more so is its moral and sociological significance for a free society. The reason is, no doubt, that the ethical universe in which ownership has its place is hard for social rationalism even to understand, let alone to find congenial. And since social rationalism is in ascendancy everywhere, it is not surprising that the institution of ownership has been badly shaken. Even discussions on questions concerning the management of firms are often conducted in terms which suggest that the owner has followed the consumer and the taxpayer into the limbo of "forgotten men." The true role of ownership can be appreciated only if we look upon it as representative of something far beyond what is visible and measurable. Ownership illustrates the fact that the market economy is a form of economic order belonging to a particular philosophy of life and to a particular social and moral universe. This we now have to define, and in so doing the word "bourgeois"* imposes itself, however much mass public opinion (especially of the intellectual masses) may, after a century of deformation by Marxist propaganda, dislike this designation or find it ridiculous.

In all honesty, we have to admit that the market economy has a bourgeois foundation. This needs to be stressed all the more because the romantic and socialist reaction against everything bourgeois has, for generations past, been astonishingly successful in turning this concept into a parody of itself from which it is very difficult to get away. The market economy, and with it social and political freedom, can thrive only as a part and under the protection of a bourgeois system. This implies the existence of a society in which certain fundamentals are respected and color the whole network of social relationships: individual effort and responsibility, absolute norms and values, independence based on ownership, prudence and daring, calculating and saving, responsibility for planning one's own life, proper coherence with the community, family feeling, a sense of tradition and the succession of generations combined with an open-minded view of the present and the future, proper tension between individual and community, firm moral discipline, respect for the value of money, the courage to grapple on one's own with life and its uncertainties, a sense of the natural order of things, and a firm scale of values. Whoever turns up his nose at these things

* The word "bourgeois" is here used to correspond to the German word *bürgerlich*, in a completely non-pejorative and non-political sense. As will be seen from the context, the word is used to designate a particular way of life and set of values.

98

or suspects them of being "reactionary" may in all seriousness be asked what scale of values and what ideals he intends to defend against Communism without having to borrow from it.

To say that the market economy belongs to a basically bourgeois total order implies that it presupposes a society which is the opposite of proletarianized society, in the wide and pregnant sense which it is my continual endeavor to explain, and also the opposite of mass society as discussed in the preceding chapter. Independence, ownership, individual reserves, saving, the sense of responsibility, rational planning of one's own life—all that is alien, if not repulsive, to proletarianized mass society. Yet precisely that is the condition of a society which cherishes its liberty. We have arrived at a point where we are simply forced to recognize that here is the true watershed between social philosophies and that every one of us must choose for himself, knowing that the choice is between irreconcilable alternatives and that the destiny of our society is at stake.

Once we have recognized this necessity of a fundamental choice, we must apply it in practice and draw the conclusions in all fields. It may come as a shock to many of us to realize how much we have already submitted to the habits of thought of an essentially unbourgeois world. This is true, not least, of economists, who like to think in terms of money flows and income flows and who are so fascinated by the mathematical elegance of fashionable macroeconomic models, by the problems of moving aggregates, by the seductions of grandiose projects for balanced growth, by the dynamizing effects of advertising or consumer credit, by the merits of "functional" public finance, or by the glamor of progress surrounding giant concerns—who are so fascinated by all this, I repeat—that they forget to consider the implications for the values and institutions of the bourgeois world, for or against which we have to decide. It is no accident that Keynes—and nobody is more responsible for this tendency among economists than he—has reaped fame and admiration for his equally banal and cynical observation that

99

"in the long run, we are all dead." And yet it should have been obvious that this remark is of the same decidedly unbourgeois spirit as the motto of the *ancien régime: Après nous le déluge.* It reveals an utterly unbourgeois unconcern for the future, which has become the mark of a certain style of modern economic policy and inveigles us into regarding it as a virtue to contract debts and as foolishness to save.

A most instructive example is the modern attitude toward an institution whose extraordinary development has caused it to become a much-discussed problem. I have in mind installment buying, or consumer credit. In its present form as a mass habit and in its extreme extent, it is certainly a conspicuous expression of an "unbourgeois" way of life. It is significant, however, that this view and the misgivings deriving from it are hardly listened to nowadays, let alone accepted. It is not, as we are often told, mere "bourgeois" prejudice but the lesson of millennial experience and consonant with man's nature and dignity and with the conditions of a sound society to regard it as an essential part of a reasonable and responsible way of life not to live from hand to mouth, to restrain impatience, self-indulgence, and improvidence alike, to think of the morrow, not to live beyond one's means, to provide for the vicissitudes of life, to try to balance income and expenditure, and to live one's life as a consistent and coherent whole extending beyond death to one's descendants rather than as a series of brief moments of enjoyment followed by the headaches of the morning after. To depart conspicuously from these precepts has always and everywhere been censured by sound societies as shiftless, spendthrift, and disreputable and has carried the odium of living as a parasite, of being incompetent and irresponsible. Even so happy-go-lucky a man as Horace was of one mind on this subject with Dickens' Mr. Micawber: "Annual income twenty pounds, annual expenditure nineteen nineteen and six, result happiness. Annual income twenty pounds, annual expenditure twenty pounds ought and six, result misery."

100

Installment buying as a mass habit practiced with increasing carelessness is contrary to the standards of the bourgeois world in which the market economy must be rooted, and jeopardizes it. It is, at the same time, an indicator of how much of the humus of the bourgeois existence and way of life has already been washed away by social erosion, as well as an infallible measure of proletarianization, not in the sense of the material standard of living but of a style of life and moral attitude. The representatives of this style of life and moral attitude have lost their roots and steadfastness; they no longer rest secure within themselves; they have, as it were, been removed from the social fabric of family and the succession of generations. They suffer, unconsciously, from inner non-fulfillment, their life as a whole is stunted, they lack the genuine and essentially non-material conditions of simple human happiness. Their existence is empty, and they try to fill this emptiness somehow. One way to escape this tantalizing emptiness is, as we have seen, intoxication with political and social ideologies, passions, and myths, and this is where Communism still finds its greatest opportunity. Another way is to chase after material gratifications, and the place of ideologies is then taken by motor scooters, television sets, by quickly acquired but unpaid-for dresses—in other words, by the flight into unabashed, immediate, and unrestrained enjoyment. To the extent that such enjoyment is balanced not only by corresponding work but also by a reasonable plan of life, saving and provision for the future, and by the non-material values of habits and attitudes transcending the moment's enjoyment, to that extent the emptiness, and with it the "unbourgeois" distress, is, in fact, overcome. But unless this is the case, enjoyment remains a deceptive method of filling the void and is no cure.[2]

The incomprehension, and even hostility, with which such reflections are usually met nowadays is one more proof of the predominance of social rationalism, with all its variants and offshoots, and of the implied threat to the foundations of the market economy. One of these offshoots is the ideal of earning a maximum amount of money in a minimum of working time and then finding an outlet in maximum consumption, facilitated by installment buying, of the standardized merchandise of modern mass production. *Homo sapiens consumens* loses sight of everything that goes to make up human happiness apart from money income and its transformation into goods. Two of the important factors that count in this context are how people work and how they spend their life outside work. Do people regard the whole of the working part of their life as a liability, or can they extract some satisfaction from it? And how do they live outside work, what do they do, what do they think, what part have they in natural, human existence? It is a false anthropology, one that lacks wisdom, misunderstands man, and distorts the concept of man, if it blinds us to the danger that material prosperity may cause the level of simple happiness not to rise but to fall because the two above-mentioned vital factors are in an unsatisfactory condition. Such anthropology also prevents us from recognizing the true nature of proletarianism and the true task of social policy.

It is, for instance, a superficial and purely materialist view of proletarianism to believe complacently that in the industrialized countries of the West the proletarians are becoming extinct like the dodo simply because of a shorter working week and higher wages, wider consumption, more effective legal protection of labor and more generous social services, and because of other achievements of current social policy. It is true that the proletariat, as understood by this kind of social rationalism, is receding. But there remains the question of whether, concurrently with this satisfactory development and perhaps because of it, ever wider classes are not engulfed in proletarianism as understood in a much more subtle sense, in the sense, that is, of a social humanism using other criteria which are really decisive for the happiness of man and the health of society. The criteria I have in mind are those which we know well already, the criteria beyond the market, beyond money incomes and their consumption. Only in the light of those criteria can we assess the tasks of genuine social policy, which I advocated fifteen years ago in my book on *The Social Crisis of Our Time* and for which Alexander Rüstow has recently coined the felicitous term of "vital policy."

The circle of our argument closes. It is, again, private ownership which principally distinguishes a non-proletarian form of life from a proletarian one. Once this is recognized, the social rationalism of our time has really been left behind. We shall see in a later chapter how direct and short a road leads from here to the great problem of our era's constant inflationary pressure, which has developed into a danger to the market economy plain for all to see.[3]

The Spiritual and Moral Setting

One of the oversimplifications by which social rationalism distorts the truth is that Communism is a weed particular to the marshes of poverty and capable of being eradicated by an improvement in the standard of living. This is a fatal misconception. Surely everyone must realize by now that the world war against Communism cannot be won with radio sets, refrigerators, and wide-screen films. It is not a contest for a better supply of goods—unfortunately for the free world, whose record in this field cannot be beaten. The truth is that it is a profound, all-encompassing conflict of two ethical systems in the widest sense, a struggle for the very conditions of man's spiritual and moral existence. Not for one moment may the free world waver in its conviction that the real danger of Communism, more terrible than the hydrogen bomb, is its threat to wipe these conditions from the face of the earth. Anyone who rejects this ultimate, apocalyptic perspective must be very careful, lest, sooner or later, and perhaps for no worse reason than weakness or ignorance, he betray the greatest and highest values which mankind has ever had to defend. In comparison with this, everything else counts as nothing.

If we want to be steadfast in this struggle, it is high time to bethink ourselves of the ethical foundations of our own economic

103

system. To this end, we need a combination of supreme moral sensitivity and economic knowledge. Economically ignorant moralism is as objectionable as morally callous economism. Ethics and economics are two equally difficult subjects, and while the former needs discerning and expert reason, the latter cannot do without humane values.[4]

Let us begin with a few questions which we, as economists, may well put to ourselves. Are we always certain of our calling? Are we never beset by the sneaking doubt that although the sphere of human thought and action with which we deal is one of primary necessity, it may, for that very reason, be of a somewhat inferior nature? *Primum vivere, deinde philosophari*—certainly. But does this dictum not reflect an order of precedence? And when the Gospel says than man does not live by bread alone, does this not imply an admonition that once his prayer for his daily bread is fulfilled, man should direct his thoughts to higher things? Should we be free of such scruples and doubts—and this is not a matter for pride—others will assuredly bring them to our attention.

I myself had a characteristic experience in this respect. Some years before his death, I had the privilege of a discussion with Benedetto Croce, one of the greatest minds of our age. I had put forward the proposition that any society, in all its aspects, is always a unit in which the separate parts are interdependent and make up a whole which cannot be put together by arbitrary choice. I had maintained that this proposition, which is now widely known and hardly challenged, applied also to the economic order, which must be understood as part of the total order of society and must correspond to the political and spiritual order. We are not free, I argued, to combine just any kind of economic order, say, a collectivist one, with any kind of political and spiritual order, in this case the liberal. Since liberty was indivisible, we could not have political and spiritual liberty without also choosing liberty in the economic field and rejecting the necessarily unfree collectivist economic order; conversely, we had to be clear in our minds that a collectivist economic order meant the destruction of political and spiritual liberty. Therefore, the economy was the front line of the defense of liberty and of all its consequences for the moral and humane pattern of our civilization. My conclusion was that to economists, above all, fell the task, both arduous and honorable, of fighting for freedom, personality, the rule of law, and the ethics of liberty at the most vulnerable part of the front. Economists, I said, had to direct their best efforts to the thorny problem of how, in the aggravating circumstances of modern industrial society, an essentially free economic order can nevertheless survive and how it can constantly be protected against the incursions or infiltrations of collectivism.

This was my part of the argument on that occasion, during the last war. Croce's astonishing reply was that there was no necessary connection between political and spiritual freedom on the one hand and economic freedom on the other. Only the first mattered; economic freedom belonged to a lower and independent sphere where we could decide at will. In the economic sphere, the only question was one of expediency in the manner of organizing our economic life, and this question was not to be related with the decisive and incomparably higher question of political and spiritual freedom. The economic question was of no concern to the philosopher, who could be liberal in the spiritual and political field and yet collectivist in the economic. The important movement for the defense of spiritual and political freedom was *liberalismo*, as Croce called it, to distinguish it from *liberismo*, by which slightening term he designated the defense of economic freedom.[5]

Croce's view hardly needs to be refuted today, and even his followers will not be inclined to defend it. But Croce's error has had a fatal influence on the development of Italian intellectuals and has smoothed the way to Communism for many of them. The mere fact that so eminent a thinker could be so utterly wrong about the place of economic matters in society proves how necessary it is to thresh this question out over and over again.

105

Naturally nobody would dream of denying that the aspect of society with which the economist deals belongs to the world of means, as opposed to ends, and that its motives and purposes therefore belong to a level which is bound to be low, if only because it is basic and at the foundation of the whole structure. This much we must grant a man like Croce. To take a drastic example, what interests economics is not the noble beauty of a medieval cathedral and the religious idea it embodies, but the worldly and matter-of-fact question of what place these monuments of religion and beauty occupied in the overall economy of their age. It is the complex of questions which, for instance, Pierre du Colombier has discussed in his charming book *Les chantiers des cathédrales*. We are fully aware that what concerns us as economists is, as it were, the prosaic and bare reverse side of the *décor*. When the materialistic interpretation of history regards the spiritual and political life of nations as a mere ideological superstructure on the material conditions of production, we are, as economists, very sensitive to the damning revelations of a philosophy of history that reduces higher to lower—a feeling which proves our unerring sense of the genuine scale of values.

All of this is so obvious that we need not waste another word on it. But equally obvious is the argument with which we must safeguard a proper place in the spiritual and moral world for the economy, which is our sphere of knowledge. What overweening arrogance there is in the disparagement of things economic, what ignorant neglect of the sum of work, sacrifice, devotion, pioneering spirit, common decency, and conscientiousness upon which depends the bare life of the world's enormous and ever growing population! The sum of all these humble things supports the whole edifice of our civilization, and without them there could be neither freedom nor justice, the masses would not have a life fit for human beings, and no helping hand would be extended to anyone. We are tempted to say what Hans Sachs angrily calls out to Walter von Stolzing in the last act of *Die Meistersinger:* "Do not despise the masters!"

106

We are all the more entitled to do so if, steering the proper middle course, we guard against exaggeration in the opposite direction. Romanticizing and moralistic contempt of the economy, including contempt of the impulses which move the market economy and the institutions which support it, must be as far from our minds as economism, materialism, and utilitarianism.

When we say *economism,* we mean one of the forms of social rationalism, which we have already met. We mean the incorrigible mania of making the means the end, of thinking only of bread and never of those other things of which the Gospel speaks. It is economism to succumb to those aberrations of social rationalism of which we have spoken and to all the implied distortions of perspective. It is economism to dismiss, as Schumpeter does, the problem of giant industrial concerns and monopoly with the highly questionable argument that mass production, the promotion of research, and the investment of monopoly profits raise the supply of goods, and to neglect to include in the calculation of these potential gains in the supply of material goods the possible losses of a non-material kind, in the form of impairment of the higher purposes of life and society. It is economism to allow material gain to obscure the danger that we may forfeit liberty, variety, and justice and that the concentration of power may grow, and it is also economism to forget that people do not live by cheaper vacuum cleaners alone but by other and higher things which may wither in the shadow of giant industries and monopolies. To take one example among many, nowhere are the economies of scale larger than in the newspaper industry, and if only a few press lords survive, they can certainly sell a maximum of printed paper at a minimum of pennies or cents; but surely the question arises of what there is to read in these papers and what such an accumulation of power signifies for freedom and culture. It is economism, we continue, to oppose local government, federalism, or decentralization of broadcasting with the argument that concentration is cheaper. It is economism, again, to measure the peasant's life exclusively by his money income without asking what else determines his existence beyond supply and demand, beyond the prices of hogs and the length of his working day; and the worst economism is the peasant's own. It is, finally, that selfsame economism which misleads us into regarding the problem of economic stability merely as one of full employment, to be safeguarded by credit and fiscal policy, forgetting that besides equilibrium of national aggregates, equal importance should be attached to the greatest possible stability of the individual's existence—just as the springs of a car are as important for smooth driving as the condition of the road.[6]

When we say *materialism*, we mean an attitude which misleads us into directing the full weight of our thought, endeavor, and action towards the satisfaction of sensual wants. Almost indissolubly linked therewith is *utilitarianism*, which, ever since the heyday of that philosophy, has been vitiating our standards in a fatal manner and still regrettably distorts the true scale of values. One of the more likable of the high priests of that cult, Macaulay, wrote in his famous essay on Francis Bacon, the ancestor of utilitarianism and pragmatism, that the production of shoes was more useful than a philosophical treatise by Seneca; but once more we must ask the familiar question of whether shoes—not to mention the latest products of progress—are likely to be of much help to a man who, in the midst of a world devoted to that cult, has lost the moral bearings of his existence and who therefore, though he may not know why, is unhappy and frustrated. It is indeed our misfortune that mankind has, but for a small remnant, dissipated and scattered the combined spiritual patrimony of Christendom and antiquity, to which Seneca contributed a more than negligible portion. This is what our reaction should be today to a passage in another and no less famous of Macaulay's essays, bursting with derision and indignation about Southey, who, at the dawn of British industry, had had the temerity to say that a cottage with rosebushes beside the door was more beautiful than the bleak workers' houses which were sprouting all over the place—"naked, and in a row."[7]

Economism, materialism, and utilitarianism have in our time merged into a cult of productivity, material expansion, and the standard of living. This cult proves once again the evil nature of the absolute, the unlimited, and the excessive. Not too long ago, André Siegfried recalled Pascal's dictum that man's dignity resided in thought, and Siegfried added that although this had been true for three thousand years and was still valid for a small European elite, the real opinion of our age was quite different. It is that man's dignity resides in the standard of living. No astute observer can fail to note that this opinion has developed into a cult, though not many people would, perhaps, now speak as frankly as C. W. Eliot, for many years President of Harvard University, who, in a commemoration address in 1909, enounced the astonishing sentence: "The Religion of the Future should concern itself with the needs of the present, with public baths, play grounds, wider and cleaner streets and better dwellings."

This cult of the standard of living scarcely needs further definition after what we have already said. It is a disorder of spiritual perception of almost pathological nature, a misjudgment of the true scale of vital values, a degradation of man not tolerable for long. It is, at the same time, very dangerous. It will, eventually, increase rather than diminish what Freud calls the discontents of civilization. The devotee of this cult is forced into a physically and psychologically ruinous and unending race with the other fellow's standard of life—keeping up with the Joneses, as they say in America—and with the income necessary for this purpose. If we stake everything on this one card and forget what really matters, freedom above all, we sacrifice more to the idol than is right, so that, if once the material standard of living should recede by an inch or fail to rise at the rate the cult demands, we remain politically and morally disarmed and baffled. We are deprived of firmness, resistance, and valor in today's world struggle, where more than the standard of life (though it, too) is at stake; we become hesitant and cowardly, until it may be too late to realize that

109

exclusive concentration on the standard of living can lose us both that standard and freedom as well. This road to happiness is bound to lead to a dead end sooner or later. As we approach the limits of reasonable consumption, the cult of the standard of life must end up in disillusionment and eventual repugnance. Even now we are told by Riesman and other American sociologists that the mass of consumers is becoming so blasé that the most spectacular advertising effort can hardly break through. Color television, the second car in the family, the television screen in the private swimming pool—all right, but what then? Fortunately, the moment seems near when people begin to rediscover the charms of books and music, of gardening and the upbringing of their children.

One thing that makes the standard-of-life cult so dangerous is that it obscures the issue in the struggle between the free world and Communism. Again and again experience has shown how grave an error it is to believe that the counterforce to Communism which must form the moral core of the defense behind the West's military and political battle lines resides in faith in the power of the standard of living. It would be foolish, of course, to belittle or deny the importance of the standard of living in this contest. But one has not understood much about the phenomenon of modern totalitarianism if one still regards as an evil fruit of poverty this infernal mixture of unbridled power and deception of the masses—with spells concocted by morally unsettled and mentally confused intellectuals.

No, the source of the poison of Communist totalitarianism is our era's social crisis as a whole, which has now spread also to the colored peoples; it is the disintegration of the social structure and its spiritual and moral foundations. Communism thrives wherever the humus of a well-founded social order and true community has been removed by proletarianization, social erosion, and the disappearance of the bourgeois and peasant classes; it thrives where men, and intellectuals above all, have lost their roots and solidity and have been pried loose from the social fabric of the family, the succession of generations, neighborliness, and other true communities. Communism finds the most fertile soil of all wherever these processes of social disintegration are associated with religious decline, as first in China and now in the Moslem world and in Japan.

Totalitarianism gains ground exactly to the extent that the human victims of this process of disintegration suffer from frustration and non-fulfillment of their life as a whole because they have lost the true, pre-eminently non-material conditions of human happiness. For this reason it is certain that the decisive battle between Communism and the free world will have to be fought, not so much on the field of material living conditions, where the victory of the West would be beyond doubt, but on the field of spiritual and moral values. Communism prospers more on empty souls than on empty stomachs. The free world will prevail only if it succeeds in filling the emptiness of the soul in its own manner and with its own values, but not with electric razors. What the free world has to set against Communism is not the cult of the standard of living and productivity or some contrary hysteria, ideology, or myth. This would merely be borrowing Communism's own weapons. What we need is to bethink ourselves quietly and soberly of truth, freedom, justice, human dignity, and respect of human life and the ultimate values. For these we must set our course unerringly; we must cherish and strengthen the spiritual and moral foundations of these values and vital goods and try to create and preserve for mankind such forms of life as are appropriate to human nature and support and protect its conditions.[8]

The material prosperity of the masses is not an absolute standard, and a warning against regarding it as the West's principal weapon in the cold war is nowhere more justified than in the underdeveloped countries. For one thing, their case makes it particularly clear that the belief that people can be preserved from Communism by higher standards of living is dangerously superficial because it grossly exaggerates a factor which in itself is not unimportant and because it forgets the decisive spiritual and moral problems. In the underdeveloped countries, another factor assumes importance. The

111

road to higher living standards is sought in industrialization, urbanization, and general emulation of the advanced Western nations' society and civilization; but, even more than in the Western world, this usually leads at once to a revolutionary upheaval in all the traditional forms of life and thought. What happens then is ominously manifest, for instance, in Japan, where the dissolution of the old forms, powerfully promoted after the last war by an obtuse victor, has prepared the ground for the Communist seed in a manner which poverty and material destruction could never have achieved. For the same reason, it is regrettable that India seems to follow Nehru's materialist socialism rather than Gandhi's humane wisdom. Finally, as regards the present advance of Communism in the Arab countries, it is unfortunately clear that it owes much less to the poverty of the masses than to the incompetence of the ruling classes, hysterical hatred of the West, and to immature intellectuals bewildered by the decline of Islam.[9]

So there is a grave danger that in the especially vulnerable field of underdeveloped countries the free world may lose, by proletarianization, urbanization, intellectualization, disintegration of family and religion, and disruption of the ancient forms of life, everything it may hope to wring from Communism by modernization, mechanization, and industrialization. There is a possibility that the non-material consequences of "economic development" may cause more losses than its material consequences cause gains, and this possibility is enhanced by the West's arrogant tendency to underestimate these nations' loyalty to their traditions. Thus we play into the hands of Communism the trump card of unnecessarily hurt national, religious, and cultural susceptibilities that are already exacerbated by a pathological feeling of inferiority vis-à-vis the Western countries. What we should do instead is to use the entirely admirable loyalty of a people to its own traditions as a bulwark against Communism; we should encourage and respect this loyalty and set its forces of preservation against the dissolving and eroding effects of material Westernization.[10]

Let us return to our main theme. Whatever may be the proper place of the economy in the universal order, what is the ethical place of the specific economic order proper to the free world? This economic order is the market economy, and it is with its relationships that economics as a science is largely concerned. What, then, are the ethical foundations of the market economy?

"Supply and demand," "profit," "competition," "interest," "free play of forces," or whatever other words we may choose to characterize the free economic system prevailing, even if in imperfect form, outside the Communist world—do they not, to say the least, belong to an ethically questionable or even reprehensible sphere? Or to put it more bluntly, are we not living in an economic world or, as R. H. Tawney says, in an "acquisitive society" which unleashes naked greed, fosters Machiavellian business methods and, indeed, allows them to become the rule, drowns all higher motives in the "icy water of egotistical calculation" (to borrow from the *Communist Manifesto*), and lets people gain the world but lose their souls? Is there any more certain way of desiccating the soul of man than the habit of constantly thinking about money and what it can buy? Is there a more potent poison than our economic system's all-pervasive commercialism? Or can we still subscribe to that astonishing eighteenth-century optimism which made Samuel Johnson say: "There are few ways in which man can be more innocently employed than in getting money"?

Economists and businessmen who have a distaste for such questions or who would, at any rate, prefer to hand them over, with a touch of irony, to theologians and philosophers are ill advised. We cannot take these questions too seriously, nor must we close our eyes to the fact that it is not necessarily the most stupid or the worst who are driven into the camp of collectivist radicalism for lack of a satisfactory answer to these questions. Among these men are many who have a right to call themselves convinced Christians.

There is another and no less important reason why we should examine the ethical content of everyday economic life. This reason

113

is that the question concerns us most intimately because it reaches down to the levels from which our roots draw their life-giving sap. *Navigare necesse est, vivere non est necesse,* says an inscription on an old sailor's house in Bremen; we may generalize this into saying: Life is not worth living if we exercise our profession only for the sake of material success and do not find in our calling an inner necessity and a meaning which transcends the mere earning of money, a meaning which gives our life dignity and strength. Whatever we do and whatever our work, we must know what place we occupy in the great edifice of society and what meaning our activity has beyond the immediate purpose of promoting material existence. We must answer to ourselves for the social functions for which society rewards us with our income. It is a petty and miserable existence that does not know this, that regards the hours devoted to work as a mere means of earning money, as a liability to be balanced only by the satisfactions which the money counterpart of work procures.[11]

This feeling for the meaning and dignity of one's profession and for the place of work in society, whatever work it be, is today lost to a shockingly large number of people. To revive this feeling is one of the most pressing tasks of our times, but it is a task whose solution requires an apt combination of economic analysis and philosophical subtlety. This is, perhaps, truer of commerce than of other callings because the merchant's functions are more difficult to place in society than others. An activity which, at first sight, seems to consist of an endless series of purchases and sales does not display its social significance and professional dignity as readily as do the peasant's or the sailor's pursuits. The merchant himself is not easily aware of them, nor are others, who all too often treat him as a mere parasite of society, an ultimately redundant intermediary whose "trading margins" are resented as an irksome levy and whom one would like to eliminate wherever possible. How infinitely more difficult must it be, then, to explain to a layman the functions of stock-exchange speculation and to defeat the almost ineradicable prejudices which fasten onto this favorite subject for anti-capitalist critics?

This is the place, too, to note that the hard-boiled business world, which ignores such questions or leaves them, with contempt, to the "unbusinesslike" intellectuals, and these same intellectuals' distrust of the business world match and mutually exacerbate each other. If the business world loses its contact with culture and the intellectuals resentfully keep their distance from economic matters, then the two spheres become irretrievably alienated from each other. We can observe this in America in the anti-intellectualism of wide circles of businessmen and the anti-capitalism of equally wide circles of intellectuals. It is true that intellectuals have infinitely less social prestige in America than in Europe and that they are much less integrated in the network of society and occupy a much more peripheral place than their brothers in Europe. They retaliate for this seating plan at the nation's table with their anti-capitalism, and the businessmen and entrepreneurs repay the intellectuals' hostility by despising them as "eggheads."

In so dynamic a competitive economy, the American intellectuals have to admit that the gulf between education and wealth, which is derided in Europe in the person of the *nouveau riche*, is the rule rather than the exception, as it should be; on the other hand, American businessmen easily fall into the habit of treating the intellectual as a pompous and would-be-clever know-all who lacks both common sense and a sound scale of values. Since in both cases the caricature is often not very far from the truth, the result is a vicious circle of mutually intensifying resentment which threatens to end up in catastrophe. One has to break out of this vicious circle by making the world of the mind as respectable to the business world as, conversely, the business world to the world of the mind.

Naturally, there is no question of taking sides with American intellectuals when they rebel against a predominantly commercial society with which they have little in common. But it must be conceded that it will not be easy to hold down this rebellion as long as

115

the tension between business and culture is not considerably diminished. This tension is particularly obvious in the United States and in all the overseas territories of European expansion. It would be unfair to expect the diminution to come from only one side, and the task would become harder if we were simply to blame the American anti-capitalist intellectuals and not try also to understand their point of view. The chain reaction between the business world's distrust of intellectuals and the intellectuals' retaliating resentment should be broken by both sides: the intellectuals should abandon untenable ideologies and theories, and the "capitalists" should adopt a philosophy which, while rendering unto the market the things that belong to the market, also renders unto the spirit what belongs to it. Both movements together should merge into a new humanism in which the market and the spirit are reconciled in common service to the highest values. It need hardly be mentioned that we Europeans have no reason to strike any holier-than-thou attitudes about these problems. If things are, on the whole, still a little better in Europe, this is due to no merit of ours but to an historical heritage which beneficially slows down a development we share with "Europe overseas."[12]

What, then, is our answer to the great question we asked at the beginning? At what ethical level, in general, must we situate the economic life of a society which puts its trust in the market economy?

It is rather like the ethical level of average man, of whom Pascal says: *"L'homme n'est ni ange ni bête, et le malheur veut que qui veut faire l'ange fait la bête."* To put it briefly, we move on an intermediate plane. It is not the summit of heroes and saints, of simon-pure altruism, selfless dedication, and contemplative calm, but neither is it the lowlands of open or concealed struggle in which force and cunning determine the victor and the vanquished.

The language of our science constantly borrows from these two contiguous spheres to describe modern economic processes, and it is characteristic of our uncertainty that we usually reach either too high or too low. When we speak of "service" to the consumer, we obviously have in mind not St. Elizabeth but the assistant who wipes the windshield of our car at the filling station, and the "conquest" of a market brings to mind the traveling salesman, tempting prospectuses, and rattling cranes rather than thundering tanks or booming naval guns.[13] It is true that in our middle plateau of everyday economic life there is, fortunately, as much room for elevations into the higher sphere of true devotion as there is for depressions of violence and fraud; nevertheless, it will generally be granted that the world in which we do business, bargain, calculate, speculate, compare bids, and explore markets ethically corresponds by and large to that middle level at which the whole of everyday life goes on. Reliance on one's own efforts, initiative under the impulse of the profit motive, the best possible satisfaction of consumer demand in order to avoid losses, safeguarding one's own interests in constant balance with the interests of others, collaboration in the guise of rivalry, solidarity, constant assessment of the weight of one's own performance on the incorruptible scales of the market, constant struggle to improve one's own real performance in order to win the prize of a better position in society— these and many other formulations are used to characterize the ethical climate of our economic world. They are imperfect, groping, and provisional, perhaps also euphemistic, but they do express what needs to be said at this point in our reflections.

This ethical climate, we must add at once, is lukewarm, without passions, without enthusiasm, but also, in the language of one of Heine's poems, without "prodigious sins" and without "crimes of blood"; it is a climate which, while not particularly nourishing for the soul, at least does not necessarily poison it. On the other hand, it is a favorable climate for a certain atmosphere of minimal consideration and for the elementary justice of a certain correspondence of give and take and most favorable, whatever one may say, for the development of productive energy. That this energy is applied not to the construction of pyramids and sumptuous palaces

117

but to the continual improvement of the well-being of the masses and that this happens because of the effect of all-powerful forces proper to the structure and ethical character of our free economic order is perhaps the greatest of the assets in its overall balance sheet.

This view of the ethical climate of the market has distinguished ancestors. In 1748, Montesquieu wrote in his book *L'Esprit des Lois* of the spirit of our market economy (which he calls *esprit de commerce*): "It creates in man a certain sense of justice, as opposed, on the one hand, to sheer robbery, but on the other also to those moral virtues which cause us not always to defend our advantage to the last and to subordinate our interest to those of others" (Book XX, Chapter 2). We may add that our era's market-economy society may claim to be less subject to compulsion and power than any other society in history, though it is perhaps for that very reason all the more prone to deception as a means of persuasion. We shall have more to say about this later.

The poem by Heine to which we alluded is "Anno 1829," and the lines we referred to are these:

> Prodigious sins I'd rather see
> And crimes of blood, enormous, grand,
> Than virtue, self-content and fat,
> Morality with cash in hand.

Who does not know such moments of despair in the face of Philistine self-satisfaction and ungenerousness? But this should not cause us to forget the real issue here, namely, the eternal romantic's contempt of the economy, a contempt shared often enough by reactionaries and revolutionaries, as well as by aloof aesthetes. Nevertheless, there remains the question of whether we really prefer to do away with "virtue" and go hungry, to give up "morality" and go bankrupt.

As a matter of fact, a certain opprobrium was attached for many centuries to that middle level of ethics which is proper to any essentially free economy. It is the merit of eighteenth-century social and moral philosophy, which is the source of our own discipline of political economy, to have liberated the crafts and commercial activities—the banausic (the Greek βάναυσος means "the man at the stove") as they were contemptuously called in the slave economy of Athens—from the stigma of the feudal era and to have obtained for them the ethical position to which they are entitled and which we now take for granted.

It was a "bourgeois" philosophy in the true sense of the word, and one might also legitimately call it "liberal." It taught us that there is nothing shameful in the self-reliance and self-assertion of the individual taking care of himself and his family, and it led us to assign their due place to the corresponding virtues of diligence, alertness, thrift, sense of duty, reliability, punctuality, and reasonableness. We have learned to regard the individual, with his family, relying on his own efforts and making his own way, as a source of vital impulses, as a life-giving creative force without which our modern world and our whole civilization are unthinkable.

In order to appreciate just how important this "bourgeois" spirit is for our world, let us consider the difficulty of implanting modern economic forms in the underdeveloped countries, which often lack the spiritual and moral conditions here under discussion. We in the West take them for granted and are therefore hardly aware of them, but the spokesmen of the underdeveloped countries frequently see only the outward economic success of Western nations and not the spiritual and moral foundations upon which it rests. A sort of human humus must be there, or at least be expected to form, if Western industry is to be successfully transplanted. Its ultimate conditions remain accuracy, reliability, a sense of time and duty, application, and that general sense of good workmanship which is obviously at home in only a few countries. With some slight exaggeration, one might put it this way: modern economic activity can thrive only where whoever says "tomorrow" means tomorrow and not some undefined time in the future.[14]

119

In the Western world, "interested" activity has, without doubt, a positive value as the mainspring of society, civilization, and culture. Some may still protest in the name of Christian teaching, but in so doing, they merely reveal that they have, for their part, not yet overcome the eschatological communism of the Acts of the Apostles.[15] After all, "the doctrine of self-reliance and self-denial, which is the foundation of political economy, was written as legibly in the New Testament as in the *Wealth of Nations*," and Lord Acton, the distinguished English historian to whom we owe this bold statement, rightly adds that this was not realized until our age.[16] The history of literature is very revealing: for Molière, the bourgeois was still a comic figure, and when for once Shakespeare introduces a merchant as such, it is Shylock. It is a long way to Goethe's *Wilhelm Meister*, where we move in the bourgeois trading world and where even double-entry bookkeeping is transfigured by philosophy and poetry.

To make this even clearer, let us turn the tables and see what happens when we give free rein to those who condemn the market, competition, profit, and self-interest in the name of a "higher" morality and who deplore the absence of the odor of sanctity in individual self-assertion. They clearly do violence to one side of human nature, a side which is essential to life and which balances the other, nobler side of selfless dedication. This kind of moralism asks too much of ordinary people and expects them constantly to deny their own interests. The first result is that the powerful motive forces of self-interest are lost to society. Secondly, the purposes of this "higher" economic morality can be made to prevail only by doing something eminently immoral, namely, by compelling people—by force or cunning and deception—to act against their own nature. In all countries in which a collectivist system has been set up, in the name of many high-sounding purposes and not least of an allegedly "higher" morality, police and penalties enforce compliance with economic commands, or else people are kept in a state of permanent intoxication by emotional ideologies and rousing propaganda—as far and as long as it may be possible.

This, as we all know, regularly happens whenever the market is replaced by a collectivist economy. The market economy has the ability to use the motive power of individual self-interest for turning the turbines of production; but if the collectivist economy is to function, it needs heroes or saints, and since there are none, it leads straight to the police state. Any attempt to base an economic order on a morality considerably higher than the common man's must end up in compulsion and the organized intoxication of the masses through propaganda. To cite Pascal again, ". . . *et qui veut faire l'ange fait la bête.*" This is one of the principal reasons for the fact, with which we are already familiar, that a free state and society presuppose a free economy. Collectivist economy, on the other hand, leads to impoverishment and tyranny, and this consequence is obviously the very opposite of "moral." Nothing could more strikingly demonstrate the positive value of self-interested action than that its denial destroys civilization and enslaves men. In "capitalism" we have a freedom of moral choice, and no one is *forced* to be a scoundrel. But this is precisely what we are forced to be in a collectivist social and economic system. It is tragically paradoxical that this should be so, but it is, because the satanic rationale of the system presses us into the service of the state machine and forces us to act against our consciences.

However, to reduce the motives of economic action solely to the desire to obtain material advantage and avoid material loss would result in too dark a picture of the ethical basis of our free economic system. The ordinary man is not such a *homo œconomicus,* just as he is neither hero nor saint. The motives which drive people toward economic success are as varied as the human soul itself. Profit and power do move people, but so do the satisfactions of professional accomplishment, the wish for recognition, the urge to improve one's performance, the dream of excavating Troy (as in the famous example of Schliemann), the impulse to help and to give, the passion of the art or book collector, and many other things.[17] But even if we discover nothing better than the motive of bare material advantage, we should never forget that the man who decently provides

121

for himself and his family by his own effort and on his own responsibility is doing no small or mean thing. It should be stated emphatically that he is more deserving of respect than those who, in
the name of a supposedly higher social morality, would leave such
provision to others. This applies also to that further category of
people who pride themselves on their generosity at others' expense
and shed tears of emotion about themselves when their advocacy
of a well-oiled welfare state earns them a place in the hearts of the
unsuspecting public—and, at the same time, on some political
party's list of candidates.

Anyone who knows anything about economics will realize at
once that these considerations suggest a familiar answer to an obvious question. What will happen when these individualist motives
induce people to do things which are manifestly harmful to others?

Again we turn to the social philosophy of the eighteenth century
and its lessons. An economy resting on division of labor, exchange,
and competition is an institution which, in spite of its occasionally
highly provocative imperfections, does tend, more than any other
economic system, to adjust the activities governed by individual
interests to the interests of the whole community. We know the
mechanism of this adjustment. The individual is forced by competition to seek his own success in serving the market, that is, the
consumer. Obedience to the market ruled by free prices is rewarded
by profit, just as disobedience is punished by loss and eventual
bankruptcy. The profits and losses of economic activity, calculated
as precisely and correctly as possible by the methods of business
economics, are thus at the same time the indispensable guide to a
rational economy as a whole. Collectivist economies, of whatever
degree of collectivism, try in vain to replace this guidance by
planning.

These simplified formulations are, of course, highly inadequate,
although the truth they contain is undeniable. We need not waste
many words over this or over the large and perhaps increasing
number of cases where even the market and competition fail to dis-

charge the enormous task of adjusting individual economic action to the common interest. It hardly needs to be stressed, either, how difficult it is to keep competition as such free and satisfactory. Any more or less well-informed person knows that these unsolved tasks and difficulties constitute the thorny problems of an active economic and social policy and that they cannot be taken too seriously.

However, this is not the place to discuss them. There is something else, though, which does need stressing in this context. Have we said all there is to say when we have underlined the importance of competition, and of the price mechanism moved by competition, in regulating an economic system whose principle it is to leave individual forces free? Is it enough to appeal to people's "enlightened self-interest" to make them realize that they serve their own best advantage by submitting to the discipline of the market and of competition?

The answer is decidedly in the negative. And at this point we emphatically draw a dividing line between ourselves and the nineteenth-century liberal utilitarianism and immanentism, whose traces are still with us. Indeed, there is a school which we can hardly call by any other name but liberal anarchism, if we reflect that its adherents seem to think that market, competition, and economic rationality provide a sufficient answer to the question of the ethical foundations of our economic system.

What is the truth? The truth is that what we have said about the forces tending to establish a middle level of ethics in our economic system applies only on the tacit assumption of a modicum of primary ethical behavior. We have made it abundantly clear that we will have no truck with a sort of economically ignorant moralism which, like Mephistopheles in reverse, always wills the good and works the bad. But we must add that we equally repudiate morally callous economism, which is insensitive to the conditions and limits that must qualify our trust in the intrinsic morality of the market economy. Once again, we must state that the market economy is not enough.

123

In other words, economic life naturally does not go on in a moral vacuum. It is constantly in danger of straying from the ethical middle level unless it is buttressed by strong moral supports. These must simply be there and, what is more, must constantly be impregnated against rot. Otherwise our free economic system and, with it, any free state and society must ultimately collapse.

This also applies in the narrower sense of competition alone. Competition is essential in restraining and channeling self-interest, but it must constantly be protected against anything tending to vitiate it, restrict it, and cause its degeneration. This cannot be done unless everybody not only accepts the concept of free and fair competition but in practice lives up to his faith. All individuals and groups, not excluding trade-unions (as must be stressed in view of a widespread social priggishness), who take part in economic life must make a constant moral effort of self-discipline, leaving as little as possible to an otherwise indispensable government-imposed compulsory discipline. It is by no means enough to invoke the laws of the market in appealing to people's enlightened self-interest and their economic reason, for within certain limits, cartels, labor unions, pressure groups, and trade associations serve their members' interests very well indeed when they exercise monopoly power or pressure on the government's economic policy in an attempt to get more than genuine and fair competition would give them. There must be higher ethical values which we can invoke successfully: justice, public spirit, kindness, and good will.[18]

So we see that even the prosaic world of business draws on ethical reserves by which it stands and falls and which are more important than economic laws and principles. Extra-economic, moral, and social integration is always a prerequisite of economic integration, on the national as on the international plane. As regards the latter, it should be especially emphasized that the true and ultimate foundation of international trade, a foundation of which our textbooks have little to say, is that unwritten code of normal ethical behavior which is epitomized in the words *pacta sunt servanda*.[19]

The market, competition, and the play of supply and demand do not create these ethical reserves; they presuppose them and consume them. These reserves have to come from outside the market, and no textbook on economics can replace them. J. B. Say was mistaken in his youthful work *Olbie ou Essai sur les moyens de réformer les moeurs d'une nation,* a liberal utopian fantasy published in 1800, when he naïvely proposed to hand the citizens of his paradise "un bon traité d'économie politique" as a "premier livre de morale." That valiant utilitarian Cobden also seems to have thought in all seriousness that free-trade theory was the best way to peace.

Self-discipline, a sense of justice, honesty, fairness, chivalry, moderation, public spirit, respect for human dignity, firm ethical norms—all of these are things which people must possess before they go to market and compete with each other. These are the indispensable supports which preserve both market and competition from degeneration. Family, church, genuine communities, and tradition are their sources. It is also necessary that people should grow up in conditions which favor such moral convictions, conditions of a natural order, conditions promoting co-operation, respecting tradition, and giving moral support to the individual. Ownership and reserves, and a feeling for both, are essential parts of such an order. We have, a little earlier, characterized such an order as "bourgeois" in the broadest sense, and it is the foundation upon which the ethics of the market economy must rest. It is an order which fosters individual independence and responsibility as much as the public spirit which connects the individual with the community and limits his greed.

The market economy is a constantly renewed texture of more or less short-lived contractual relations. It can, therefore, have no permanence unless the confidence which any contract presupposes rests on a broad and solid ethical base in all market parties. It depends upon a satisfactory average degree of personal integrity and, at the margin, upon a system of law which counteracts the natural

125

tendency to slip back into less-than-average integrity. Within that legal framework, the market's own sanctions undeniably foster the habit of observing certain minimum rules of behavior and thereby also integrity. Whoever always lies and deceives and breaks contracts will sooner or later be taught that honesty is the best policy. For all its resting on utilitarian calculation, this pattern of behavior is valuable and reliable, as we can see in the extreme example of Soviet Russia, which, in its relations with the outside world of the market, has tried systematically and successfully to acquire the reputation for prompt payment while adhering, in other respects, to the ethical code of gangsters. Even if we conscientiously credit the market with certain educational influences, we are, therefore, led back to our main contention that the ultimate moral support of the market economy lies outside the market. Market and competition are far from generating their moral prerequisites autonomously. This is the error of liberal immanentism. These prerequisites must be furnished from outside, and it is, on the contrary, the market and competition which constantly strain them, draw upon them, and consume them.

We would, of course, again err on the side of unrealistic and unhistorical moralism if we were to apply to modern economic behavior moral standards which would have been enough to condemn mankind at any time because men can never live up to them. Such moralism is least tolerable when it self-righteously pretends that the moralist is a better man for the mere reason that his standards are so strict. This should always be remembered whenever the talk turns to the questionable aspects of competition. Ruthless rivalry has never and in no circumstances been banned from human society. The young Torrigiani, spurred by jealousy and professional rivalry, smashed Michelangelo's nasal bone and thereby disfigured him for life; in our days, a leading German trade-union intellectual, no doubt a valiant detractor of the "capitalist jungle," tried to get rid of a rival by means of forged letters—it is always the same thing and always equally unedifying. But we get nowhere by raising our eyebrows because the market economy does not always display the sporting spirit of a tennis tournament; we would do better to reflect that no small advantage of the market economy is that it channels men's natural rivalry into forms which, by and large, are preferable to broken noses and forged letters—and also to mass executions, as in Communist countries.

But we cannot, in good conscience, let the matter rest there. It cannot be denied that the market places the constant competitive struggle for self-assertion and self-advancement in the center of the stage. Nor can it be denied that such all-pervasive competition has a disturbing tendency to lead to consequences to which we cannot remain indifferent, especially from the moral point of view. Those who are in the rough-and-tumble of the competition of modern economic life, with its nerve-racking claims on time, effort, and susceptibility, and who are worn down by this endless struggle are more sensitive than most to the questions raised thereby, and it would be both unjust and uncivil therefore to treat them as monopoly-mongers.

We all acknowledge the validity and justice of such questions when we accept as a model of a higher form of rivalry the way in which certain professions, above all the medical, submit to strict rules of competition to the point of including them among the standards of professional behavior. Unfortunately, this example of the medical profession's deontology cannot be applied to industry and trade. But it shows what a blessing for all it would be if a definite code of competitive behavior, resting on professional standards, binding for all and violable only at the price of outlawry, were to dampen competition everywhere and withdraw it from the laws of "marginal ethics,"[20] without appeal to the state but in full appreciation of the positive potentialities of professional solidarity.

In acknowledging these potentialities, we express the idea that we should aim at compensating the socially disintegrating effects of competition by the integrating forces outside the market and outside competition. There is, however, the danger of abuse. On no account must competition be corrupted by its economically most questionable and morally most reprehensible perversion, namely, monopoly in any shape or form. Monopoly is precisely the worst form of that commercialism which we want to combat by trying to mitigate competition by integrating counterforces.

The truth is that competition, which we need as a regulator in a free economy, comes up on all sides against limits which we would not wish it to transgress. It remains morally and socially dangerous and can be defended only up to a point and with qualifications and modifications of all kinds. A spirit of ever alert and suspicious rivalry, not too particular in the choice of its means, must not be allowed to predominate and to sway society in all its spheres, or it will poison men's souls, destroy civilization, and ultimately disintegrate the economy.

To assert oneself all the time by ubiquitous advertising, day and night, in town and country, on the air and on every free square foot of wall space, in prose and in verse, in word and picture, by open assault or by the subtler means of "public relations," until every gesture of courtesy, kindness, and neighborliness is degraded into a move behind which we suspect ulterior motives; to fashion all imaginable relations and performances on the principle of supply and demand and so to commercialize them, not excluding art and science and religion; forever to compare one's own position with that of others; always to try out something new, to shift from one profession and from one place to the next; to look with constant jealousy and envy upon others—such extreme commercialization, restlessness, and rivalry are an infallible way of destroying the free economy by morally blind exaggeration of its principle. This is bound to end up in an unhealthy state of which the worst must be feared.

The curse of commercialization is that it results in the standards of the market spreading into regions which should remain beyond supply and demand. This vitiates the true purposes, dignity, and savor of life and thereby makes it unbearably ugly, undignified, and dull. We have had occasion earlier to note this. Think of Mother's Day, a day set aside to honor mothers and motherhood; the most tender and sacred human relationship is turned into a means of sales promotion by advertising experts and made to turn the wheels of business. Father's Day soon followed, and if we did not fortunately know better, the latest forms of Christmas might make us suspect that this whip which makes the top of business spin is also a creation of modern advertising techniques. Not long ago it happened that an automobile race, which, to the horror of the spectators, led to a fatal accident, was nevertheless continued because of its commercial and technical purposes, so that even death had to defer to business and technology.

All of this cannot be castigated too severely—with the intention, not of condemning the market economy, but of stressing the need to circumscribe and moderate it and of showing once more its dependence upon moral reserves. This circumscription and moderation can take many forms. One of them is that we do not allow competition to become the dominating principle and that we keep an eye on all the circumstances which tend to mitigate it. Let me illustrate my point. Has any sociologist ever bothered to discover why there is usually fierce rivalry among actors and singers, while circus folk tend to live in an atmosphere of kindly good-fellowship? Would it not be a rewarding task to examine the whole texture of modern society for such differences in competition and their presumable causes?[21]

Nobilitas Naturalis

It cannot be said often enough that in the last resort competition has to be circumscribed and mitigated by moral forces within the market parties. These constitute the true "countervailing power" of which the American economist J. K. Galbraith speaks in his book of the same title, and not the mechanics of organized buying power, to which he mistakenly looks for the containment of competition

129

and its monopolistic perversions. Without a fund of effective convictions regarding the moral limits of competition, the problem cannot find a genuine solution.

In a sound society, leadership, responsibility, and exemplary defense of the society's guiding norms and values must be the exalted duty and unchallengeable right of a minority that forms and is willingly and respectfully recognized as the apex of a social pyramid hierarchically structured by performance. Mass society, such as we have described it earlier, must be counteracted by individual leadership—not on the part of original geniuses or eccentrics or will-o'-the-wisp intellectuals, but, on the contrary, on the part of people with courage to reject eccentric novelty for the sake of the "old truths" which Goethe admonishes us to hold on to and for the sake of historically proved, indestructible, and simple human values. In other words, we need the leadership of genuine *clercs* or of men such as those whom the distinguished psychiatrist Joachim Bodamer recently described as "ascetics of civilization," secularized saints as it were, who in our age occupy a place which must not for long remain vacant at any time and in any society. That is what those have in mind who say that the "revolt of the masses" must be countered by another revolt, the "revolt of the elite."

The conviction is rightly gaining ground that the important thing is that every society should have a small but influential group of leaders who feel themselves to be the whole community's guardians of inviolable norms and values and who strictly live up to this guardianship. What we need is true *nobilitas naturalis*. No era can do without it, least of all ours, when so much is shaking and crumbling away. We need a natural nobility whose authority is, fortunately, readily accepted by all men, an elite deriving its title solely from supreme performance and peerless moral example and invested with the moral dignity of such a life. Only a few from every stratum of society can ascend into this thin layer of natural nobility. The way to it is an exemplary and slowly maturing life of dedicated endeavor on behalf of all, unimpeachable integrity, constant restraint of our common greed, proved soundness of judgment, a spotless private life, indomitable courage in standing up for truth and law, and generally the highest example. This is how the few, carried upward by the trust of the people, gradually attain to a position above the classes, interests, passions, wickedness, and foolishness of men and finally become the nation's conscience. To belong to this group of moral aristocrats should be the highest and most desirable aim, next to which all the other triumphs of life are pale and insipid.

No free society, least of all ours, which threatens to degenerate into mass society, can subsist without such a class of censors. The continued existence of our free world will ultimately depend on whether our age can produce a sufficient number of such aristocrats of public spirit, aristocrats of a kind which was by no means rare in the feudal age. We need businessmen, farmers, and bankers who view the great questions of economic policy unprejudiced by their own immediate and short-run economic interests; trade-union leaders who realize that they share with the president of the national bank the responsibility for the country's currency; journalists who resist the temptation to flatter mass tastes or to succumb to political passions and court cheap success and instead guide public opinion with moderation, sound judgment, and a high sense of responsibility. In turn, it will be of crucial importance for the ultimate fate of the market economy whether this aristocracy includes, above all, people who, by position and conviction, have close ties with the market economy and who feel responsible for it in the moral sphere here under discussion.[22]

Evidently, many and sometimes difficult conditions must be fulfilled if such a natural aristocracy is to develop and endure and if it is to discharge its tasks. It must grow and mature, and the slowness of its ripening is matched by the swiftness of its possible destruction. Wealth gained and lost overnight is a stony ground on which it cannot prosper but on which thrive plutocracy and newly rich parvenus—the very opposite of what is desirable. Yet without

131

wealth and its inheritance, whereby a spiritual and moral tradition is handed down together with its material foundation, a natural aristocracy is equally impossible, and it would be shortsighted egalitarian radicalism to overlook this.[23] One generation is often, indeed usually, not sufficient to produce the flower and fruit of aristocratic public spirit and leadership, and this is why the almost confiscatory limitation of the testator's rights, which today is the rule in some major Western countries, is one of the most harmful measures imaginable and contrary to the spirit of sound policy.

But *richesse oblige*. Any privilege, be it a privilege of birth, mind, honor and respect, or of wealth, confers rights only in exactly the same measure in which it is accepted as an obligation. It will not do to hide one's talent in the earth; each must remain conscious of the responsibilities which his privileged position entails. If ever the much-abused words "social justice" are appropriate, it is here.

One of the obligations of wealth, which need not be enumerated, is to contribute to the filling of the gaps left by the market because they are in the realm of goods outside the play of supply and demand, but which gaps must not be left for the state to fill if we want to preserve a free society. I have in mind the patronage of art in the widest sense, generous grants for theatre, opera, music, the visual arts, and science—briefly, for everything whose existence and development would be jeopardized if it had to "pay." We would be hard put to name a single supreme work of art in any period of history which did not owe its origin to patronage, and it is even more difficult to think of a theatre, opera house, or orchestra which bowed to the laws of supply and demand without damage to its quality or which, therefore, could have maintained its quality without patronage. The tragedies of Aeschylus, Sophocles, and Euripides are as unthinkable without the public donations of the rich Athenians as are the plays of Shakespeare without his patrons. Conversely, in so far as in our age the laws of supply and demand determine the level of artistic performance—in extreme form, in

132

the film industry—the devastating effects are plain for all to see.

This function is to be fulfilled by the rich in the same spirit in which in the old days the Hanseatic burghers of Bremen used to pay property taxes: in honest self-assessment of one's ability to pay and in voluntary fulfillment of an honorary duty.[24] Here it is appropriate to emphasize that this spirit is smothered by the modern welfare state and its fiscal socialism. It may also be pointed out that the rich cannot exercise their function of patronage of the arts unless they are at home in the realm of the spirit and of beauty as much as in the world of business—which brings us back to what we said earlier in this chapter.

The task of leadership falls to the natural aristocracy by virtue of an unwritten but therefore no less valid right which is indistinguishable from duty. Washington's successor, the great American statesman John Adams, had some very pertinent things to say about this. According to him, a member of the "natural aristocracy of virtues and talents" was anyone who disposed not only of his own vote but, at the same time, of the votes of those whose opinions he influenced by his example, acknowledged authority, and persuasion. But since this is unfortunately true not only of the "natural aristocracy of virtues and talents" but of everybody who, by foul means or fair, influences the formation of political opinion, we must add the qualification that the unwritten plural franchise which actually exists in any democracy is the more justified the more we can rely on the existence and effectiveness of a genuine natural aristocracy. The latter therefore appears all the more indispensable.

Finally, we have to speak of science, whose leadership functions and responsibility are obvious. There can be no doubt that here, too, rights and duties are inextricably linked. Here, too, authority— and it is authority of the highest rank—has to be gained and held by achievement and character. But what, precisely, is the deontology of science, especially, in this context, of the social sciences?

Boswell has recorded an apposite remark by Samuel Johnson, that great eighteenth-century Englishman. Certain professions,

133

Johnson said, principally the sailor's and the soldier's, had the dignity of danger. "Mankind reverence those who have got over fear, which is so general a weakness." Conversely, the honor of those professions whose dignity is danger cannot be more deeply wounded than by casting doubt on their courage.

The esteem in which science is held certainly does not rest on such a dignity of danger. We do not expect of a Sanskrit scholar the bravery of a soldier or sailor who, professionally, has to face physical danger of losing his life, but we do expect men of science to be courageous and intrepid in another sense, which we recognize when we have grasped that the "dignity of science" is truth. This sounds a little pompous, but it is meant to express something very simple. It does not mean that science is respected because it has to offer "truth" like ripe plums. What we mean is this: just as much as fear, another universal human failing is a tendency to allow the prospect of advantages or the threat of disadvantages to deflect one from the pursuit of the "true" facts and, even more, from the free announcement of facts recognized as "true." The dignity of science is that its genuine apostles constantly have to overcome this human weakness of interested squinting at truth. Only those who fulfill this requirement can partake in the dignity of science. Only they discharge the obligation put upon them by the privilege of being the servants of science, and only they can hope to attain to natural nobility and to render to the community those services which it has a right to expect from them.

Since men of science, too, are generally neither saints nor heroes, it is no doubt hard for them to live up to this standard without faltering and occasional aberrations. It is hardest for those who, unlike the Sanskrit scholar, have chosen a field of knowledge which gives them occasion and indeed obliges them to defend the "dignity of truth" in the rough-and-tumble of interests and passions. Economic policy, of which we are treating here, is such a battlefield, and the scholars involved are the jurists and economists.[25] Economists also have this in common with jurists: their scientific author-

134

ity, whose moral foundation is the "dignity of truth," is appealed to in controversial questions. Such activities by scholars are as old as the history of science and the universities; we have but to remember that in the fourteenth century Louis of Bavaria called on the famed scholars of the Universities of Bologna and Paris for opinions in his struggle with Pope John XXII. Curiously enough, this is not usually held against jurists, although the delicate nature of such a task is obvious. It always presents the man of science with a question of conscience which he must decide in the light of the "dignity of truth."

The answer should not be in doubt. Such a commission can be accepted—and indeed has point for the questioner—only if it is discharged in such a manner that the scholar's answer does not deviate in the slightest respect from that which he would have pronounced without the commission and without the ensuing advantages (which may include such things as enhanced prestige or public honors). The answer must be strictly in line with his scientific convictions, and if there is the slightest doubt about this, the scholar should withdraw. The economist, in particular, should make it a rule to put his scientific work at the service of any precise commission, originating from the government or from international or non-governmental organizations, only on condition that this work can serve his own convictions also and on the further condition that he may hope thereby to promote a good cause threatened by overwhelming forces. In the absence of these conditions, the economist has every reason to ask himself whether the counsel expected of him in the struggle of economic interests and social passions is not a mortgage on his conscience, considering the social function of his science.

If a task so undertaken also happens to involve some private interests, the economist can congratulate himself. Aims of economic policy which lack such a solid anchor have little prospect of being taken seriously in our world of overwhelming material interests and stormy passions. To take an important example, liberal trade policy would be in a bad way indeed if there did not, fortunately, exist groups which have a material interest in it and thereby form a natural counterweight to the fatal combination of protectionist interests and political passions such as nationalism and socialism. To help such groups may be regarded as a legitimate duty by the economist who weighs the opposing forces against each other.

The economist has all the less right to evade the duty of bringing his authority to bear on the controversies of economic policy since this duty has an important characteristic in common with every genuine duty. This is that it tends to be beset by vexations, and to withstand these vexations requires that same courage which is indispensable for defending the "dignity of truth." By putting his view onto one of the scales, he lessens the relative weight of the other, and the interest and passions involved on the other side will feel provoked. They have a perfect right to resist by trying to prove that the reasons, assumptions, and conclusions of the inconvenient scientific verdict are wrong and that the scientific judgment against them is a misjudgment. The scholar would be foolish if he thought himself in possession of objective truth, and it is no dishonor for him to be disproved. But he has a right to expect that his search for truth, his intellectual integrity, is not suspected. Like the judge, he has an absolute claim, which should be effectively protected, to the assurance that factual criticism of his sentence will not be replaced by an attempt to smear his reputation with accusations of bribery, cowardice, or political prejudice.

Bad experiences of this kind do not seem to have been spared even Adam Smith, the father of economics and contemporary of Samuel Johnson. In a famous passage of his *Wealth of Nations* (Book IV, Chapter 2), he says that anyone who opposes unconquerable private interests or has authority enough to be able to thwart them must expect that "neither the most acknowledged probity, nor the highest rank, nor the greatest public services can protect him from the most infamous abuse and detraction, from personal insults, nor sometimes from real danger."

136

The Asymmetry of the Market Economy

The role of natural nobility in general, and of science in particular, is seen with special clarity if we consider a very important circumstance which often does not receive sufficient attention. I have in mind what we might call the asymmetry of the market.

We know well enough that it would be foolish to regard the market, competition, and the play of supply and demand as institutions of which we can always expect the best in all circumstances. Nobody has better reasons to bear this in mind than the friend of the market economy. This general recognition leads us to a more particular one. The market frequently weights the scales in vital questions because it favors activities which are the source of gain and does not give sufficient scope to reasons which oppose these activities and should, in the general interest, have the greatest weight. The market thereby loses its authority in the ultimately most important decisions. It would be shortsighted of us to invoke the market or rely on it in such cases, and it becomes inevitable that we should seek decisions outside the market, beyond supply and demand. It is precisely for this purpose that the weight of authoritative opinions is needed. The highest interests of the community and the indispensable things of life have no exchange value and are neglected if supply and demand are allowed to dominate the field. We shall illustrate this point with a few particularly important examples.

The first example is advertising, a matter which repeatedly demands attention because it separates our era from all earlier ones as little else does, so much so that we might well call our century the age of advertising. A vast industry with enormous turnover figures lives on advertising, and it has generated such a colossus of influence and vested interests that it is hard to raise one's voice against it except in a book, all other instruments of public opinion having moved so close to the colossus that, to say the least, they can no longer be regarded as free agents.

137

We do not have to be told that advertising fulfills indispensable functions. Far be it from us to inveigh against it.[26] But only the blind could fail to notice that commercialism, that is, the luxuriance of the market and its principles, causes the beauty of the landscape and the harmony of cities to be sacrificed to advertising. The reason that the danger is so great is that although money can be made from advertising, it cannot be made from resistance to advertising's excesses and perversions. Thousands get hard cash out of advertising, but the unsalable beauty and harmony of a country give to all a sense of well-being which cannot be measured by the market. Yet the non-marketable value, while incomparably higher than the marketable one, is bound to lose unless we come to its assistance and put on its scale enough moral weight to make up for the deficiency of mercantile weight. The market's asymmetry opens a gap which has to be closed from without, from beyond the market, and it would be sheer suicide on the part of the market economy's friends to leave to others the cheap triumph of this discovery. In one of the loveliest parts of Germany there lives an old man who has dedicated his life to fighting against the excesses of advertising; it is a downright desperate fight against ignorance, greed, and obtuseness, and he carries on this struggle out of love for beauty and harmony and out of devotion. This old man is a living embodiment of our proposition that the market economy is not enough and, at the same time, proof that it needs such wise and public-spirited men as much as competition and the free play of supply and demand.

Another closely related example is installment buying, of which we have already spoken in another context. Again, there is no symmetry in the market economy between the forces favoring this extraordinarily widespread modern form of sales promotion and the forces which impede it. Yet the warmest supporter of installment buying will not deny that it is in danger of excess and degeneration. As in the first case, the asymmetry is due to the fact that the impulses originating in the market work to the benefit of consumer credit because the interests of those who want to sell their wares are joined by the special interests of the finance institutes making money out of installment-plan sales. But no money is to be made by organizing cash purchases because they need no organization. Nevertheless, not to make debts is the sound practice and the one which should primarily be encouraged. So the cause of reasonable conduct, which is threatened from all sides, needs our support and encouragement, and we should do well to reinforce the brakes, which are none too strong. We can count ourselves fortunate that this almost abandoned cause still finds some active supporters in a few economic groups, such as the savings banks and isolated industrial and trading companies, whose own interests seem to lie there. Even so, cash trade will remain in a bad enough way, thanks to the above-mentioned asymmetry of the market.[27]

One last example. the free world's trade with the Communist countries, euphemistically called East-West trade.[28] Here we meet a familiar state of affairs. This trade is highly dangerous and objectionable and is apt to strengthen the power which the free world, if it is not to delude itself, must recognize as its own worst enemy and which, indeed, never misses an opportunity of stating this with brutal frankness or of making it clear by its attitude. But money can be made only by expanding East-West trade, not by restricting it. We have a paradoxical situation: on the one side, Moscow is anxious to make good the deficiencies of the Communist economic system by getting supplies of the most wanted goods from the market economies of the free world while, at the same time, plotting these economies' destruction; on the other side, Moscow has no stauncher allies in these designs than the Western businessmen, precisely the people who represent an economic system that is the diametric opposite of Communism and who would be the first to be eliminated if Communism were to win.

The cultural and political ideal for which the West fights and the defense of which is the meaning of its struggle against Communism is the ideal of freedom in the precise sense that politics

139

must not encroach upon the whole of life and society but must leave a large part of them independent. In other words, the West opposes its own pluralistic system to Communism's monolithic one. This is the pride and the strength of the West and one of the essential conditions of the world of freedom in which alone we can breathe. The freedom of society resides in its pluralism and is defined by it; and one of the areas which must remain independent is, of course, the economy. By contrast, it is of the essence of the Communist empire that its economy and also its economic relations, as well as its cultural and all other relations with the Western world, are subordinated to the paramount purposes of politics.

We are faced with a totalitarian world empire which draws all matters, and above all the economy, into politics. It follows that each and every economic transaction with the Communist empire is an act of international politics, for the simple reason that the other party regards it as such. For this reason, any appeal to separate East-West trade discussions from politics reveals either unusual ignorance or an intention to further Communist aims, for it admirably suits Moscow's game to represent the matter as harmless. It is a weakness of the West that the decisively political character of East-West trade is easy to obscure by invoking the principle of pluralistic liberty. For monolithic Communism, trade with the West is primarily a political act: for the pluralistic West, it is primarily an opportunity for business and profit. It is precisely the habit of respecting business interests which leads Western politicians to lend their ears to businessmen who profit from East-West trade and want to transpose into this political mine field the functions of business which are legitimate and proved in our own economic and social order. There are but few who stop to think whether in this case their business interests are not in conflict with overall political interests and with political interests, at that, which are a matter of life and death for all of us and most of all for Western "capitalists."

The fact that the wind of private business interests fills the sails of the Western business world's eagerness to expand East-West trade is no proof that it has political reason on its side—and political reason must, here, have the last word. On the contrary, just because these private interests are strong, any attempts at justifying East-West trade need to be scrutinized with the greatest suspicion. Market and profit are not competent in the decision; the decision lies with higher political interests and business must submit to them. Businessmen should really regard it as an insult to their intelligence when Moscow tries to catch them with the bait of profit. They should remember Lenin's statement that when it was time to hang the world's capitalists, they would trip over each other in their eagerness to sell the Communists the necessary ropes. Unless they are completely blinded by their short-term interests, Western businessmen should not find it so very difficult to see through Moscow's dishonest game. They should realize that this is another case of asymmetry in the market, one to be stressed especially by the market's friends.

Seeing that we are, here, up against one of the limits of the market economy, it is, perhaps, hardly to be expected that businessmen themselves will exercise self-restraint for the sake of higher political interests, especially since competition works against such self-restraint; but we certainly have a right to expect that any restrictions imposed by the government in the exercise of its proper functions will be recognized as necessary, reasonable, and binding. The supporters of the market economy do it the worst service by not observing its limits and conditions, as clear in this case as in the others, and by not drawing the necessary conclusions.

The Political Framework of the Market Economy

What happens if governments, in this as in other instances, fail to take independent decisions based on objective assessment of all relevant facts and designed to serve the common interest? What if governments give way to pressures for another decision?

These questions touch upon a very sore spot. It is one to which we cannot pay too much attention in this field beyond supply and demand. To put it briefly, the problem is whether, in a mass democracy, with its many kinds of perversions, it is at all possible for policy to serve the common interest. In effect, policy has to withstand not only the pressure of powerful interest groups but also mass opinions, mass emotions, and mass passions that are guided, inflamed, and exploited by pressure groups, demagogy, and party machines alike. All these influences are more dangerous than ever when the decisions in question, to be reasonable, require unusual factual knowledge and the just assessment of all circumstances and interests involved. This applies above all to the wide field of economic policy.[29]

Of these influences, we shall first single out interest groups. We shall have to be careful not to throw out the baby with the bath water. Such groups had no place in the original concept of the modern democratic state. The idea was that there was no room for legitimate separate interests beside what was called the common interest. The state was supposed to represent an indivisible common interest through co-operation between the executive, organized in the civil service, and parliamentary parties, which, in their turn, were to be divided by ideas rather than by material interests.[30] It is well known that actual developments were less and less in line with this concept. Governments and political parties everywhere progressively became subject to the influence of groups and associations either pressing their particular claims upon both the legislative body and the administration or at least obstructing what did not suit them. One result is that political parties are swayed more by interests than by ideas; another, that the internal authority of the state and its claim to represent the common interest are impaired.

Thus the monistic state of democratic doctrine has developed into the pluralistic state of democratic practice. Although the written constitution proclaims the theory, it is complemented by the un-

142

written paraconstitutional influence of particular groups embodied in vast mass organizations and interest groupings, in powerful concerns and cartels, in farmers' unions and labor unions. The Capitol is besieged by pressure groups, lobbyists, and veto groups, to use the American political jargon. The structure of the modern state is the result of this interplay of constitutional institutions and paraconstitutional economic and social power. It is obvious that the discrepancy between democratic idea and constitutional law on the one hand and the hard facts of reality on the other puts a heavy strain on the modern democratic state. The idea itself appears compromised, and any responsible government must examine carefully all the possible means of resisting this pluralistic disintegration of the state. This process has accompanied the development of the modern state since its origins; more than a hundred years ago, Benjamin Constant, the great theoretician of constitutional government, warned against its dangers.[31] But it was only in the last quarter of the nineteenth century that it gained conspicuously in extent and pace, and in our day it has reached a degree critical for democracy and for rational economic policy. No legislative act, no import duty, no important administrative measure escapes the attention of the pressure groups and their frequently successful attempts to deflect the government's action to their own advantage.

It would be preaching to the converted to inveigh against the dangers of this development, but a few dispassionate remarks may be all the more useful.

The first circumstance which should give us cause for reflection is that the expression "pluralism," which is here used in a derogatory sense, has a positive meaning in the Anglo-Saxon countries and has been used by ourselves more than once in that meaning. In this positive sense it implies something which is a source of pride and satisfaction: the salutary existence of counterweights to the overweening power of the democratic doctrine's monistic state, the *république une et indivisible*. Has not Montesquieu, too, spoken of the *corps intermédiaires*, whose necessary function it is to loosen

143

the giant unity of the state by geographical or professional separatism? Is it not our own conviction that the centralist monistic
state is to be rejected? Is it not one of the distinguishing marks of
a sound state to allow as much social, political, and intellectual independence as possible and to leave room for local government and
autonomy, institutions and corporations, private groups with particular interests and particular rights? Is this not desirable in order
to contain the state's own striving for power, especially the democratic state's, which is all the more dangerous for posing as the
representative of the "will of the people"? Are we, then, not en-
tangling ourselves in a grave contradiction when we criticize
"pluralism"?

The contradiction is resolved if we distinguish two kinds of
pluralism, one justified and one unjustified, one sound and one
unhealthy.

By sound pluralism we mean the case of particular groups defending themselves and their rights against the power of the state
and the claims of other groups represented therein. This is a salutary limitation. A clear case in point is the landlords' effort to prevent themselves, a politically weak minority, from being expropriated by the votes of the politically strong majority of tenants.
Unhealthy pluralism, on the other hand, is not defensive but offensive. It does not limit the power of the state but tries to use it for
its own purposes and make it subservient to these purposes. The
state is opposed only when it crosses the interests of this kind of
pluralism, which, for the rest, merely tries to exploit its power.

The immense danger of this unhealthy pluralism is that pressure
groups covetously beset the state—the modern suitors of Penelope.
The wider the limits of the state's competence and the greater its
power, the more interesting it becomes as an object of desire. The
fewer the groups sharing the booty, the better it is for the partici-
pants in the marauding expedition. The ideal of such pluralism
would be to maximize the power of the state in the economy and to

minimize the number of those competing for the conquest and exploitation of that power. This ideal is achieved in the collectivist state, with the important difference, however, that it is usual in this case for an entirely new power group to triumph, which cheats all others of the booty.

These characteristics of unhealthy (offensive) pluralism explain why, during the last thirty or forty years, it has gained ground in exactly the same measure in which liberal economic policy has been displaced by centralist socialist policies. In the same measure, too, the opposite, defensive and sound pluralism, which we welcome, has lost influence and weight. State power on the one hand and economic and social power on the other have grown continuously and have progressively merged. The counterweights against this accumulation and alliance of power are federalism, local government, family, market economy, ownership, private enterprise, well-earned rights, *corps intermédiaires*—but they have become ever lighter during that period and by virtue of the same development.

If we want to understand fully the nefarious effects of offensive interest groups, we must consider what I have called "pluralism of the second degree" *(The Social Crisis of Our Time,* p. 131). By this I mean that the mass organizations of interested parties dangerously increase the already alarming power of separate interests, to the detriment of the common interest. Moreover, the representation of these interests tends to stray into dubious paths because the officials of these organizations make a living from the representation of interests and therefore have a particular concern to justify their profession as ostentatiously as possible. They therefore not only tend to be more ruthless than those whose interests it is their business to defend, but they are constantly tempted to do so in a manner which demonstrates the useful and indispensable nature of their office as clearly as possible. It is obvious that this professional vested interest of the representatives of particular interests tends to interpret the latter in the light of the former and that the two

145

need not necessarily coincide. There is a refraction of the interests represented as they pass through the prism of the officials' own particular interest.

The matter can be illustrated by an example which is of paramount importance today, namely, trade-unions. The prime interest of trade-union leaders is a continuous rise in money wages because this is a tangible and patent result of their efforts; they generally have only a secondary interest in raising real wages through price reductions or in other purposes which, for the well-being and happiness of workers and employees, may well be more important than wage increases. It is quite possible for price reductions to further the true interests of trade-union members better than wage increases, but from the point of view of the trade-union leaders themselves, price reductions have the disadvantage of obscuring their own merits. We shall see presently that this is undoubtedly one of the chief sources of the permanent inflation which characterizes the Western world today and also the reason why a "labour standard," as Hicks says, has come to replace the old gold standard—though not at all to our benefit.[32]

So much for the power of pressure groups. If we now add the power of mass opinions, mass emotions, and mass passions, the combined effect of these forces and influences on economic policy will hardly seem surprising. A first result is that economic policy will tend to be irrational, that is, determined by what is "politically feasible" rather than by what is economically rational and just. The most spectacular example is that rent control, an irrational, ill-considered, and at the same time unsocial and inequitable intervention if ever there was one, can carry the day against unexceptionable arguments and against the better judgment of honest and intelligent politicians. Rent control is really nothing but the protection of one privileged special kind of tenants, those with old leases, at the expense of the landlords and later tenants alike. Yet it persists, and the explanation is no doubt that, on the one hand, it does need a little reflection and intelligence to see its full implications and that, on the other hand, politicians are afraid to renounce this object of cheap demagogy.[33]

A second result is that the power of pressure groups and the power of mass opinions, emotions, and passions mutually support each other and that group interests can be furthered by exploiting and mobilizing the ignorance, thoughtlessness, and vague feelings of the masses. This leads to the third result: that economic policy suffers from contradictions and degenerates into a sum of disconnected measures lacking a consistent principle. A telling example of the ensuing makeshift opportunism is that of a French finance minister who recently attacked, not the causes of inflation, but only its statistics, namely, the cost-of-living index, which determines the wage level in France. Where there are no principles or where principles cannot be effectively implemented, economic policy is at the mercy of the day's political whims and so becomes a dangerous source of uncertainty, which merely aggravates nervousness and vacillation. All of this together is bound to impart to economic policy one overriding quality: it will follow political expediency, the line of least social resistance, the motto *après nous le déluge* (or, to quote Keynes again: "In the long run, we are all dead.")

This means that contemporary economic policy tends to prefer what Walter Lippmann calls "soft" solutions, solutions which appear cheapest and most convenient at the moment, even if at the expense of the future. One of these is protectionism; to bar inconvenient foreign competition is often the solution which comes to mind first, among other reasons because it is, politically, the easiest. A second type of "soft" solution is reliance on the public treasury, powerfully supported by the "fiscalism" of our times. This reliance, incidentally, is, like the demand for protective tariffs and other import restrictions, nourished by the people's obstinate inclination to believe in a sort of "fourth dimension" of economic and social policy and to forget that someone has to foot the bill, in one case the consumer, in the other the taxpayer. The consumer and the taxpayer become the "forgotten men" of our age—together with the saver and the other victims of the erosion of the value of money. The third "soft" solution, as everybody knows, is inflation—and the "softer" for starting more mildly. This is the real key to the Western countries' chronic inflation, which therefore, for reasons still to be discussed, deserves the name of "democratic-social inflation."

This diagnosis must be pronounced with ruthless honesty because recognition of the danger is the first condition of overcoming it and also because this is the best service that can be rendered to a democracy threatened by its own excesses. The danger can be countered only by a long-term and comprehensive program.

A solution must be found to the problem of how the executive can gain in strength and independence so that it can become the safeguard of continuity and common interest without curtailing the essentials of democracy, namely, the dependence of government upon the consent of those governed, which alone makes government legitimate, and without giving rise to bureaucratic arbitrariness and omnipotence. It is urgently necessary to strengthen the feeling for the imponderable nature of community surpassing all separate interests and immediate claims and commanding the individual's loyalty, even unto death; it is equally necessary to strengthen the feeling for the unchallengeable authority and power of government legitimately entrusted with managing the affairs of the community. At the same time, however, people must be liberated from the fear—only too justified in our days—of being at the mercy of a Leviathan. It is an enormously difficult problem. There can be no solution unless the state's overgrown functions are drastically pruned and its economic, financial, and social policies are once more made subject to firm, simple, and universally understood rules inspired by the common interest and by a free economic order, without which there can be no protection against arbitrary power.

The most important aspect is, again, the spiritual and moral one. Individualism and utilitarianism, which give the individual's inter-

148

ests and material profit so damaging a predominance, and legal positivism, which sees no further than the written law, must be counterbalanced by all the imponderables which ultimately are the basis of the nation as a permanent entity and without which disintegration is inevitable: the immutable standards of natural law, continuity, tradition, historical awareness, love of country, all the things which anchor a community in the hearts of men. The younger the state is and the more provisional it appears, the more pressingly must all efforts be directed toward this aim.

To this end, it is invaluable to have independent institutions beyond the arena of conflicts of interests—institutions possessing the authority of guardians of universal and lasting values which cannot be bought. I have in mind the judiciary, the central bank, the churches, universities, and foundations, a few newspapers and periodicals of unimpeachable integrity, an educational system which, by cultivating the universal and the classical, sets up a barrier to the teachings of utilitarianism and the specialization of knowledge, and, finally, that natural nobility of which we have already spoken.

In conclusion, I want to say a little more about the tasks and responsibilities falling to the academic representatives of economics in an age in which the conditions of rational economic policy serving the interests of the community and of a free society are more than ever threatened by the forces of mass democracy. Some people seem to think that the principal function of economics is to prepare the domination of society by "specialists" in economics, statistics, and planning, that is, a situation which I propose to describe as economocracy—a horrible word for a horrible thing. We have already gone quite far along this path, although it is no less dangerous to deliver state and society into the hands of such economists than into the hands of generals.[34]

The true task of economics appears to me to be quite different, especially in a modern mass democracy. Its unglamorous but all the more useful mission is to make the logic of things heard in the midst of the passions and interests of public life, to bring to light inconvenient facts and relationships, to weigh everything and assign it its due place, to prick bubbles and expose illusions and confusions, and to counter political enthusiasm and its possible aberrations with economic reason and demagogy with truth. Economics should be an anti-ideological, anti-utopian, disillusioning science. It could then render society the invaluable service of lowering the temperature of political passions, counteracting mass myths, and making life difficult for demagogues, financial wizards, and economic magicians. But economics must not itself become the willing servant of passions, of whose stultifying effects Dante says in Canto 13 of his *Paradiso: "E poi l'affetto lo intelletto lega."*

The mission of economics is understood even better if we consider a problem which is peculiar to modern democracy and keeps recurring in economic policy. I have in mind the delay between some economic or social claim and its demagogic exploitation, on the one hand, and the moment, on the other, when the price of its fulfillment can no longer be concealed. If the economist repeatedly succeeds in reducing this delay by timely and effective explanation, he renders society a service which cannot be valued too highly, for in economic policy, as elsewhere, Chateaubriand's words are true: *"Le crime n'est pas toujours puni dans ce monde; les fautes le sont toujours."*

This does not by any means imply that we economists may retire into the ivory tower of scientific neutrality. Least of all can social scientists be spared a decision at the cross-roads of our civilization; we must not only be able to read the signs, but we must know which way to point and lead: the road to freedom, humanity, and unswerving truth or the road to serfdom, violation of human nature, and falsehood. To evade this decision would be just as much *trahison des clercs* as to sacrifice the dignity of our science, which is truth, to the political and social passions of our time.

150

CHAPTER IV

Welfare State and Chronic Inflation

Communism is no immediate danger to the countries of the free Western world, nor does the specter of totalitarianism rear its ugly head among us, however great may be the threat of slow internal corruption and unscrupulous attack from outside. Neither a fully planned economy and general socialization nor the totalitarian state which necessarily goes with both are purposes for which the broad masses of the electorate can be successfully roused. What threatens the structure of our economy and society from within is something else: chronic diseases, spreading secretly and thereby all the more malignant. Their causes are hard to discover and their true nature is concealed from the superficial or thoughtless observer; they tempt individuals and groups with immediate advantages, while their fatal consequences take a long time to manifest themselves and are widely dispersed. This is precisely why these diseases are so greatly to be feared.

Among these slowly spreading cancers of our Western economy and society, two stand out: the apparently irresistible advance of the welfare state and the erosion of the value of money, which is called creeping inflation. There is a close link between the two through their common causes and mutual reinforcement. Both start slowly, but after a while the pace quickens until the deterioration is hard to arrest, and this multiplies the danger. If people knew what awaits them at the end, they would perhaps stop in good time. But the trouble is—and here we link up directly with the preceding chapter—that it is extraordinarily difficult to make the voice of reason heard while there is still time. Social demagogues use the promises of the welfare state and inflationary policy to seduce the masses, and it is hard to warn people convincingly of the price ultimately to be paid by all. All the more reason is there for those who take a more sober and longer view to redouble their efforts to undeceive the others, regardless of violent attacks from social demagogues, who are none too particular in their choice of means, and from the officials of the welfare state itself.

Another characteristic common to the welfare state and chronic inflation is that both show, clearly and alarmingly, how the political forces referred to in the preceding chapter undermine the foundations of a free and productive economy and society. Both are the outcome of mass opinions, mass claims, mass emotions, and mass passions, and both are directed by these forces against property, law, social differentiation, tradition, continuity, and common interest. Both turn the state and the ballot into means for advancing one part of the community at the expense of the others in the direction in which the majority of voters push by means of their sheer weight. Both are an expression of the dissolution of firm moral principles which were formerly accepted as self-evident.

Limits and Dangers of the Welfare State

There are, however, considerable differences between the welfare state and chronic inflation. Against inflation, the only proper attitude is one of resolute and indignant rejection; the slightest qualification of this attitude is wrong. But the concept of the welfare state encompasses much that cannot simply be rejected out of hand. Our concern, therefore, is not simply to condemn the welfare state as such but to determine its limits and dangers. We must observe the maxim put forward in the preceding chapter, namely, that the economist who is anxious to live up to his responsibilities must be careful which side he supports.

There can be no doubt that the time when the welfare state stood in need of our assistance and advocacy has passed. There is no likelihood that the indispensable minimum of government-organized security will be lacking in this era of mass democracy, robust social powers, unleashed egalitarianism, and almost habitual "robbery by the ballot." On the other hand, it is very likely indeed that this minimum may be dangerously exceeded, to the detriment of the people, the health of society, and the strength of our economy. There need be no hesitation, therefore, about which side we should support with whatever strength we may possess. It is the limits and dangers of the welfare state which require our critical attention, rather than its increasingly doubtful blessings.

A remarkable change has certainly taken place in all countries since 1945. The words "Beveridge Plan" should suffice to recall the time, more than a decade ago, when many circles enthusiastically welcomed the idea which found in the Beveridge Plan its most interesting expression.[1] Laymen and experts alike thought then that the postwar future belonged to such a "welfare state." In fact, keen efforts were made everywhere, and most of all in countries exclusively or largely dominated by socialist influences, to create such a state of guaranteed security and income equalization. Additional impetus was lent to this development by mistaken forecasts, which gave rise to the fear of a great wave of unemployment after the war.

The enthusiasm has been dissipated everywhere, even in Great Britain and the Scandinavian countries. The ideal of the welfare state has given way to its everyday practice. Disillusionment and disappointment, even misgivings and bitterness, are spreading, and critical voices are raised which are not to be ignored.[2] Few people can still close their eyes to the contrast between the extraordinary successes of a social and economic order relying on the regulating and stimulating forces of the market and free enterprise, on the one hand, and on the other the results of a continuous redis-

153

tribution of income and wealth for the sake of equality. It is a contrast which is intolerable in the long run. One or the other will have to yield—the free society and economy or the modern welfare state. To use the words of another distinguished British economist, Lionel Robbins, a man who weighs his words carefully, "the free society is not to be built on envy."[3]

The strange thing is that this bloated welfare state of ours is really an anachronism. Organized public assistance for the economically weak originated and had significance in a definite period of economic and social history, the period between the pre-industrial and today's advanced industrial society, when the old social pattern dissolved and the individual, deprived of its support, became a helpless proletarian. Thus a vacuum was created and there was a need for relief and assistance which could hardly have been met adequately without public funds, private charity notwithstanding. The paradox is that today the modern welfare state carries to an excess the system of government-organized mass relief precisely at a moment when the economically advanced countries have largely emerged from that transition period and when, therefore, the potentialities of voluntary self-help by the individual or group are greatly enhanced.

Government-organized relief for the masses is simply the crutch of a society crippled by proletarianism, an expedient adapted to the economic and moral immaturity of the classes which emerged from the decomposition of the old social order. This expedient was necessary as long as most factory workers were too poor to help themselves, too paralyzed by their proletarian position to be provident, and too disconnected from the old social fabric to rely on the solidarity and help of genuine small communities. It can be dispensed with in the degree in which we may hope to overcome that inglorious period of proletarianization and rootlessness.

In so far, then, as the advanced countries have emerged from that phase and can count upon a normal degree of individual providence, the principle of the welfare state has outlived its necessity.

It is difficult to understand why the welfare state grows so exuberantly just now when it has lost much of its urgency. People regard as progress something which surely derives its origin and meaning from the conditions of a now all but finished transition period of economic and social development. They forget that if we are to take respect for human personality seriously, we ought, on the contrary, to measure progress by the degree to which the broad masses of the people can today be expected to provide for themselves out of their own means and on their own responsibility, through saving and insurance and the manifold forms of voluntary group aid. Only this is ultimately proper to free and mature men; they should not constantly look to the government for help which in the last resort can be paid for only out of the taxpayers' pockets or by the restrictions which the devaluation of money forces upon its victims.

Are we to call it progress if we continuously increase the number of people to be treated as economic minors and therefore to remain under the tutelage of the state? Is it not, on the contrary, progress if the broad masses of the people come of age economically, thanks to their rising incomes, and become responsible for themselves so that we can cut down the welfare state instead of inflating it more and more? If government-organized mass relief is the crutch of a society crippled by proletarianism and enmassment, then we should direct all our efforts to being able to do without this crutch. This is true progress, from whatever point of view we look at it. It can be measured by our success in steadily widening the area of individual and voluntary group providence at the expense of compulsory public providence. In the same measure we shall also overcome proletarianization and enmassment and the overriding danger of degrading man into an obedient domestic animal in the state's giant stables, into which we are being herded and more or less well fed.

The objection is sometimes raised against this viewpoint that while it is true that economic improvement has lessened the masses' needs for public help, the loosening of family ties has increased these needs. It cannot be denied that family ties have loosened. But

155

we may ask whether the masses' need for help has not been far more diminished by higher incomes than increased by the loosening of family ties. Secondly, we may observe that there is no reason at all why we should simply accept the dissolution of family and family solidarity. A short time ago, a member of the House of Commons movingly described her father's plight in order to prove how inadequate the welfare state still is. But this is no proof of the urgency of public help; it is merely an alarming sign of the disappearance of natural feelings in the welfare state. In fact, the lady in question received the only proper answer when another member of Parliament told her that she should be ashamed if her father was not adequately looked after by his own daughter.

The modern welfare state, which appears an anachronism in the light of these reflections, would be incomprehensible if we failed to consider that it has changed its meaning. Its essential purpose is no longer to help the weak and needy, whose shoulders are not strong enough for the burden of life and its vicissitudes. This purpose is receding and, indeed, frequently to the detriment of the neediest. Today's welfare state is not simply an improved version of the old institutions of social insurance and public assistance. In an increasing number of countries it has become the tool of a social revolution aiming at the greatest possible equality of income and wealth. The dominating motive is no longer compassion but envy.[4]

Taking has become at least as important as giving. In the absence of a sufficient number of genuinely needy people, they have to be invented, so that the leveling down of wealth to a normal average, which satisfies social grievances, can be justified by moralistic phrasemaking. The language of the old paternal government is still current and so are its categories, but all this is becoming a screen that hides the new crusade against anything which dares exceed the average, be it in income, wealth, or performance. The aim of this social revolution is not achieved until everything has been reduced to one level, and the remaining small differences give even greater cause for social resentment; on the other hand, it is im-

156

possible to imagine a situation in which social resentment finds nothing to fasten on any more. In these circumstances there can be no foreseeable end to this development as long as the fatuous social philosophy which underlies the modern welfare state is not recognized and rejected as one of the great errors of our time.[5] The increasingly obvious ill effects of the welfare state, which include chronic inflation, should help to bring us to our senses.

Several approaches are possible in trying to define more closely the revolutionary change of which the welfare state is an expression. We could say, for example, that it is the outcome of a three-stage development during the last one hundred years, beginning with the stage of individual relief graded according to genuine needs, passing through public social insurance, and ending up in today's stage of universal, all-encompassing security. Another interpretation is related to the first. It is that the first stage was one of assistance and was designed to be self-liquidating as soon as possible; this was followed by the idea that government help should become a permanent institution, though a selective one, to be drawn upon only in well-defined cases. The last stage is that of today's revolutionary principle, which turns the state into an income pump, working day and night, with tubes and valves, with suction and pressure flows, just as its inventor Lord Beveridge described it more than ten years ago.

Whichever way we look at it, the revolutionary character of the most recent phase of the development is obvious. A whole world divides a state which occasionally rescues some unfortunate individual from destitution from another state where, in the name of economic equality and to the accompaniment of the progressive blunting of individual responsibility, a sizable part of private income is constantly sucked into the pumping engine of the welfare state and diverted by it, with considerable friction losses. Everything into the same pot, everything out of the same pot—this is becoming the ideal. As an astute British critic sarcastically puts it: "Everything must be free and equal—except the progressive taxation out of which it is all financed." (Walter Hagenbuch, in *Lloyd's Bank Review* [July 1953], p. 16).[6]

The sound old conservative and philanthropic principle that even the poorest should have something to fall back on has changed into quite another: the spreading socialization of the use of income, resting on the leveling and state-idolizing theory that any expansion of social services for the masses is a milestone of progress. Since in this system genuine individual need, as ascertained from case to case, ceases to be the standard of relief, it so happens, as we have said, that the poorest and weakest are frequently the losers. The unmistakably collectivist character of the welfare state leads in the extreme case to what another British critic, Colm Brogan, has called the pocket-money state. It is a state which deprives people of the right to dispose freely of their income by taking it away from them in taxes and which, by compensation, and after deduction of the extraordinarily high administrative costs of the system, takes over the responsibility for the satisfaction of the more essential needs, either wholly (as in the case of education or medical care) or in part (as in the case of subsidized housing or food). What people eventually retain from their income is pocket money, to be spent on television or football pools.

A hundred years ago Heinrich Heine epitomized the ideal of an egalitarian and collectivist epicureanism in the following lines:

> O sugar peas for all the world,
> Let pods yield up their marrow;
> The heavens above we gladly leave
> To angel and to sparrow.[7]

The "sugar peas for all the world" have come true, thanks to a socialization of life such as Heine would have abhorred, notwithstanding his theoretical flirtation with socialism. But whether they make up for what Heine irreverently describes as "the heavens above" is another and very doubtful question.

158

The situation which the leading welfare-state countries have already reached, and which others are aiming at, startlingly coincides with the famous vision which Alexis de Tocqueville, Heine's contemporary, saw in his mind's eye when he described the coming state in his classical work *Democracy in America:* "[The government] covers the surface of society with a network of small complicated rules, minute and uniform, through which the most original minds and the most energetic characters cannot penetrate, to rise above the crowd. The will of man is not shattered, but softened, bent, and guided; men are seldom forced by it to act, but they are constantly restrained from acting. Such a power does not destroy, but it prevents existence; it does not tyrannize, but it compresses, enervates, extinguishes, and stupefies a people, till each nation is reduced to nothing better than a flock of timid and industrious animals, of which the government is the shepherd." (Vol. II, Book IV, Chapter 6, p. 319.) A leading German socialist recently ventured the remark (in an article in the *Deutsche Rundschau*) that, thanks to the development of the welfare state, the "humanization of the state," Pestalozzi's noble aim, was giving way, even this side of the Iron Curtain, to the "etatization of man."

So much for the revolutionary character of the modern welfare state. Its traces are ubiquitous. One of them is the apparently irresistible extension of public providence to ever wider classes who would certainly provide for themselves if left alone but are now put under the tutelage of the state. Equally striking is another peculiarity of the modern welfare state which is intimately connected with its nature. In the old days, public assistance was, as we noted, intended as a subsidiary and temporary substitute for people's own provision for themselves and as such was meant to safeguard only a certain minimum; nowadays, public services are increasingly becoming the rule, often with the hardly veiled intention of meeting maximal or, indeed, luxury standards. Nothing is, in any case, dearer to the hearts of the new ideologists of fiscal socialism than the highest possible taxation, and we can be certain that they feel no irresistible urge to economize in fields where they can confer blessings upon the broad masses of voters.

Perhaps we can make all of this even clearer if we illustrate the change with a few examples. A very fruitful field for this purpose is, again, housing policy. Almost all countries are familiar with this particular manifestation of the welfare state. The old and commendable principle that there are a few marginal problems on the housing market which justify a helping hand has been transformed beyond recognition. With the war and its consequences as a pretext, it has been replaced by a long-term policy of low rents, first at the expense of the politically weak minority of landlords, who are thus to all intents and purposes expropriated in some countries; then at the expense of the taxpayers, who, of course, largely coincide with the subsidized tenants, so that they pay in taxes what they save in rent; and then at the expense of the tenants of non-subsidized new buildings, whose rents are pushed up by the system of rent control; and finally at the expense of the nation's capital stock. We have reached the point where it seems odd even to ask why everybody should not, as used to be the rule, pay out of their own pocket the full cost price of their apartment just as they pay for their clothes.

Another very characteristic change has taken place in the equally important field of education. In many countries, the old and tested principle of helping gifted young people with scholarships, but for the rest expecting parents to make a contribution to the cost of their children's higher education, has been replaced by the ideal of a public and uniform system of education that is free at all levels and thereby completely socialized. One hardly dares put forward the notion that there is nothing wrong in expecting parents normally to make a sacrifice for their children's education. The consequences of this kind of educational Jacobinism are becoming ever more visible, and they may eventually lead to a swing in public opinion. In Great Britain, where the development has gone furthest, parents who are prepared to make personal sacrifices in order to offer their children a better education than they receive free in the state's school machine are suspected of not having the right "social" attitude. Again one might ask why it should be proper and natural to pay all the expenses of an automobile out of one's own pocket but shift the expenses onto the state, that is, onto the taxpayer and hence possibly back onto oneself, in the case of the education of one's children; but, as in other cases, the very question is heretical and a sign of reprehensible views.[8]

As a last important example, let us take the admittedly difficult one of medical services. The road from old-style social policy to the modern welfare state can again be clearly traced. The original principle that the economically weakest should be relieved of the risk of costly operations or prolonged sickness has gradually changed in our generation to something entirely different. Step by step, health services have been socialized, the British National Health Service being the summit this side of the Iron Curtain; the exception has become the rule and the assistance granted for genuine needs has been transformed into a permanent system.

In this manner we are getting further and further from the rule that people who can provide for themselves in other respects should, in principle, also provide in their private budget for sickness, relying, if they wish, on insurance as an institution invented for the risks of the unforeseeable. This should at any rate be regarded as the sound and normal principle appropriate to a market economy, and it should find the widest possible application. The situation into which compulsory health insurance has got in the majority of the Western industrial countries urgently suggests that we should remind ourselves of this principle. Compulsory health insurance itself is seriously ill nearly everywhere, and a recovery must be sought in the following principal ways: first, compulsory insurance should be limited to those classes for whom the risk of sickness constitutes a serious burden and who are not easily amenable to voluntary insurance; secondly, we should encourage all those mani-

161

fold forms of decentralized assistance for which Switzerland may be held up as a model; and thirdly, we should introduce into all systems of sickness insurance universal and sizable individual cost contributions which can easily be adjusted in cases of hardship.[9]

Let us now try to assess the significance of this welfare state for modern civilization, society, economy, and public life. Naturally, we can do no more than stress a few salient points.

We begin with a circumstance that is of particular importance in view of all the misgivings already mentioned and still to be mentioned. The dangers of the welfare state are the more serious because there is nothing in its nature to limit it from within. On the contrary, it has the opposite and very vigorous tendency to go on expanding. All the more is it necessary to impose limits from without and to keep a sharp and critical eye on it. By its continuous expansion, the welfare state tries to cover more and more uncertainties of life and ever wider circles of the population, but it also tends to increase its burdens; and the reason why this is so dangerous is that while expansion is easy and tempting, any repeal of a measure later recognized as hasty is difficult and ultimately politically unfeasible.

It is hard to imagine that Great Britain would have set up the National Health Service in its present far-reaching form if people had realized in advance how it would work out, or even if some questions, which now appear elementary, had been raised and thought through in time.[10] It is equally hard to imagine how this venture could be undone today and so people try to make the best of it. But any further step along the road to the welfare state should be considered with the utmost caution, with a very clear view of the consequences and in the knowledge that, like the reduction of the minimum voting age, it is normally irreversible.

The welfare state not only lacks automatic brakes and not only gathers impetus as it moves along, it also moves along a one-way street in which it is, to all intents and purposes, impossible or, at any rate, exceedingly difficult to turn back. What is more, this road undoubtedly leads to a situation where the center of gravity of society shifts upwards, away from genuine communities, small, human, and warm, to the center of impersonal public administration and the impersonal mass organizations flanking it. This implies growing centralization of decision and responsibility and growing collectivization of the individual's welfare and design for life.

The effects of this development should be examined carefully in all respects. So far we have been able to rely on the reactions of individuals who know that they must assume responsibility for certain risks; but we must be clear in our minds that the welfare state, by shifting the center of gravity of decision and responsibility upwards, weakens or distorts these reactions. What is the effect on production if individuals are relieved of the consequences of bad performance but at the same time also deprived of incentives for good performance, especially performance entailing some risk? What is the effect on such important decisions as those relating to saving and investment? What happens to the birth rate, which in the past was limited, to some extent, by the fact that the individual remained responsible for his own family, whatever its size, whereas now he is relieved of that responsibility or even allowed to cash in on procreation? These are some of the questions which every unprejudiced person ought to ask today.

The individual and his sense of responsibility constitute the secret mainspring of society, and this mainspring is in danger of slackening if the welfare state's leveling machine lessens both the positive effects of better performance and the negative ones of worse performance. It is not surprising that some observers, including no less a man than Field Marshal Montgomery, should begin to wonder whether the overgrown welfare state is not well on the way to undermining the moral and social health of the nation which succumbs to its temptations. Something of this kind must have been in Goethe's mind when, two years before the French Revolution, he wrote this prophetic sentence: "I must say, I believe that humanism will eventually prevail; but I am afraid that at the same time the world will become a huge hospital, with everyone nursing his neighbor." (*Italienische Reise II*, Naples, May 27, 1787.)

Nor must we pass over in silence another question, which has already been put in all seriousness and which, indeed, can hardly be evaded. It is the question of whether the crushing costs of the welfare state, which can no longer be reduced without political inconvenience, are not one of the major factors impairing the free world's resolution and the strength of its military defense against the Communist empire and thus forcing the West to concentrate more and more on nuclear defense. To be sure, this does not prevent precisely those who sympathize most with the welfare state from wanting to snatch from the West even this last desperate weapon which the welfare state has left it.

The past's extreme individualism is not least to blame for the reversal which has brought about the opposite extreme, the modern welfare state. It is surely the mark of a sound society that the center of gravity of decision and responsibility lies midway between the two extremes of individual and state, within genuine and small communities, of which the most indispensable, primary, and natural is the family. And surely it is our task to encourage the development of the great variety of small and medium communities and thereby of group assistance within circles which still have room for voluntary action, a sense of responsibility, and human contact and which avoid the cold impersonality of mass social services.

The modern welfare state is, without any doubt, an answer to the disintegration of genuine communities during the last one hundred years. This disintegration is one of the worst legacies the past has left us, whether we call it mass civilization, proletarianization, or any other name. But it is the wrong answer. I said this more than ten years ago, when it was the essence of my criticism of the Beveridge Plan. Far from curing this disease of our civilization, the welfare state alleviates a few symptoms of the disease at the cost of its gradual aggravation and eventual incurability. It is, for instance, a lamentable misunderstanding of the problem to permit the family-allowance funds to absorb into the state's income-pumping system even the family itself.

There is worse to come. If the modern state increasingly takes it upon itself to hand out welfare and security on all sides—first to the advantage of one group, then of another—it must degenerate into an institution which fosters moral disintegration and prepares its own eventual doom. We are again reminded of Frédéric Bastiat's malicious definition; the modern state fits it more and more closely. It also confirms Dean Inge, who pessimistically regarded politics as the art of conjuring money out of the pockets of the opposite party into those of one's own party and making a living thereby.

The morally edifying character of a policy which robs Peter in order to pay Paul cannot be said to be immediately obvious. But it degenerates into an absurd two-way pumping of money when the state robs nearly everybody and pays nearly everybody, so that no one knows in the end whether he has gained or lost in the game. It would also be well not to bring in morality when social grievances and ruthless pressure-group politics end up in claims to the well-earned income and property of others, and hence in the confiscatory taxation with which we have all become familiar.

It is true, of course, that people do not always realize that when they turn to the state for the fulfillment of their wishes their claims can be satisfied only at the expense of others. We have met the underlying sophism before. It rests on the habit of regarding the state as a kind of fourth dimension, without stopping to think that its till has to be filled by the taxpayers as a whole. A money claim on the state is always an indirect claim on somebody else, whose taxes contribute to the sum demanded; it is a mere transfer of purchasing power through the medium of the state and its compulsory powers. It is astonishing for how long this natural and simple fact can be obscured by the modern welfare state.

The more widely the principle of the welfare state is applied, the closer comes the moment when the giant pumping engine turns out to be a deception for everybody and becomes an end in itself, which eventually serves no one except the mechanics who make a living out of its manipulation, namely, bureaucrats. They naturally have an interest in obscuring the facts. There is, however, one circumstance which should help us to understand how this deception can be worked for so long; it is the fact that few things have contributed more to the most recent development of the welfare state than the concept, born of the Great Depression, that society was immensely rich but that its wealth remained potential as long as monetary circulation was faulty, and could be transformed into actual wealth by increasing effective demand. The wealth so liberated from its slumbers would then be justly distributed by the welfare state. At the same time—and this is one of the most popular conclusions drawn from the Keynesian doctrine—this redistribution of income would increase mass consumption and reduce saving and would thus be the best means of insuring full employment and keeping the welfare state's springs flowing.

It was the depression of the thirties which fostered this faith in a sort of self-financing of the comprehensive welfare state, another kind of "fourth dimension"; and it is this faith alone which can explain the recklessness with which the problem of the cost of the welfare state has been neglected for so long.

Today the time of illusions is past. It has become clear, and it is widely said, especially in Great Britain,[11] that if one seriously wants to put the welfare state into practice, one has to use taxation to stir up income distribution at all levels and has to draw even on the lowest income groups to help finance the cost. The burden of the system of mass social services, which the state enforces, can no longer be borne by the higher incomes alone but must be placed on the shoulders of those same masses whose interests the system is to serve. This means that to a large extent the money is conjured from people's right pocket into their left, with a detour via the treasury

166

and the enormous friction losses entailed thereby. It has become clear now that, under the spell of the "poverty amidst plenty" illusion, people overestimated the potential wealth even in the most favorable case. It has also become clear that there is a price to be paid in the form of the costs of ever more powerful state machinery, of a blunting of the will to work and of individual responsibility, and of the dreary grayness of a society in which vexation at the top and envy at the bottom choke civic sense, public spirit, creative leisure, neighborliness, generosity, and genuine community. What remains is the pumping engine of Leviathan, the insatiable modern state.

The utmost limit of the welfare state, then, lies at that point where its pumping engine begins to deceive everybody. Some nations have already reached this point. One may ask the heretical question of whether everyone would not be better off if the welfare state were dismantled, except for an indispensable minimum, and if the money thus saved were left to non-governmental forms of social services.[12] The question gains in urgency by the fact that there are legitimate doubts about whether the enormous tax burden, to which the commitments of the welfare state contribute decisively, is in the long run at all compatible with a free economic order and whether it can continue without permanent inflationary pressure.

Another very grave aspect of this development generally receives but scant attention. It is that fashionable social phraseology is apt to obscure the fact that the direct or indirect compulsion inherent in the welfare state tends to politicize social security. The consequences are obvious. Security from the risks of life is at the mercy of both the state's bureaucracy and political strife. Thus our so richly paradoxical age praises as progress that which, in fact, enhances the power of the national state. The more we appeal to the solidarity of people of the same nationality or domicile and the more we fuse them into a "national community" in which money is transferred backwards and forwards, the more perfectly shall we

167

"nationalize" man to the detriment of a free international community of peoples and their solidarity.

In the nineteenth century, Ernest Renan could still define a nation as a *"plébiscite de tous les jours"*; now we are approaching the day when we can define it as a pension fund, a compulsory insurance scheme in which passport and certificate of residence are a free insurance policy, an income pump *de tous les jours*. Saving and private insurance are forms of provision against risks which belong to the area of economic rationality, the market, private law, and freedom. They are not bounded by national frontiers. The field of private investment and insurance is the whole world; but national social security falls into the area of politics, collectivist organization, public law, and compulsion and therefore locks people in behind the bars of the national state. Social services whose backbone is the state's compulsion are, strictly speaking, national services, and social insurance is nothing but national insurance—unless, of course, we think of a world state, where Germans, Italians, Argentinians, and Ethiopians join in a world pension fund.

The list of the welfare state's paradoxes and illusions is not yet exhausted. A further circumstance deserves mention. Very many people imagine that taxation of the higher income brackets merely implies restriction of luxury spending and that the purchasing power skimmed off from above is channeled into "social" purposes down below. This is an elementary error. It is quite obvious that larger incomes (and larger wealth) have so far mainly been spent for purposes which are in the interests of all. They serve functions which society cannot do without in any circumstances. Capital formation, investment, cultural expenditure, charity, and patronage of the arts may be mentioned among many others. If a sufficient number of people are wealthy and if they are dispersed, then it is possible for a man like Alexander von Humboldt to pay out of his own pocket for scientific ventures of value to everyone or for Justus von Liebig to finance his own research. Then it is possible, too, that there should be private teachers' posts and thousands of other rungs on the ladder on which the gifted can climb and the very variety of which makes it much more likely that some help will be forthcoming somewhere, whereas in the modern welfare state their fate depends upon the decision of one single official or upon the chances of one single examination.[13]

If, then, the higher income groups are crushed by progressive taxation, it is obvious that some of their functions will have to be dropped and, since they are indispensable, taken over by the state—even if it is only the maintenance of some historic monument which used to be private property. To this extent, at any rate, the purchasing power taken from above is not at the disposal of the welfare state. It must be reserved for the purpose of paying, with public funds, for private services made impossible by taxation. This nullifies the aim of the welfare state. If the welfare state should claim any merit for educating, say, a genius like Gauss at public expense, the answer is that in the actual case of Gauss the task was discharged excellently and quite unbureaucratically, not only by the Duke of Brunswick, but also by a lot of others who would today be prevented from doing so by the welfare state's taxation or would, at any rate, be left with little incentive or inclination to spend their money in this way.

In this case, then, the upper income groups' loss of purchasing power is not matched by any gain on the part of the lower income groups. The benefit goes not to the masses but to the state, which waxes in power and influence. At the same time, a powerful stimulus is given to modern state absolutism, with its centralization of decisions on very important matters, such as capital formation, investment, education, scientific research, art, and politics. What used to be personal and voluntary service is today at best state service, centralized, impersonal, compulsory, crudely stereotyped, and bought at the price of curtailed freedom.

Inevitably, such socialization of income uses for socially important functions must make a country's moral climate oppressive. Kindliness, honorary office, generosity, quiet conversation, *otium cum dignitate,* everything which Burke calls by the now familiar name of the unbought graces of life—all of that suffocates under the stranglehold of the state. Everything—paradoxically in a welfare state—is commercialized, everything an object of calculation, everything forced through the state's money-income pump. Hardly anything is done on an honorary basis any more because few can afford it; civic sense and public spirit are transformed into vexation at the top and envy at the bottom. In these circumstances everything that is done is done professionally and for money. There is a narrower margin of income available for free gifts, voluntary sacrifice, a cultivated way of life, and a certain breadth of spending, and for this reason the climate is not congenial to munificence, diversity, good taste, community, and public spirit. Civilization is blighted.

This is one of the roots of the leaden boredom which—as we have had occasion to note earlier—seems to be a distinguishing feature of the advanced welfare state. Another root of this evil is closely connected. It is that the welfare state, contrary to its proclaimed aim, tends to petrify the economic and social stratification and may impede rather than facilitate movement between classes. Severe taxation, especially in the form of steeply progressive income taxes, must surely hit those incomes most which are high enough to allow for the accumulation of wealth and the assumption of business risks.[14] Is this not bound (for a number of other reasons, too, which this is not the place to discuss) to make it more difficult to set up new businesses and to acquire property? Does this not imply that it is becoming much harder for anybody to work himself up above the broad, low plain of propertyless income earners? And does it not also become far less attractive even to try to do so, especially since the welfare state itself takes care of a sort of comfortable stall-feeding of the domesticated masses? Is this not bound to work to the benefit precisely of existing large firms? At the same time, life in such a country becomes about as exciting and entertaining as a game of cards in which the winnings are equally divided between the partners at the end. It seems a hopeless undertaking in these circumstances to try to raise oneself economically or socially, unless one chooses to go into administration, whether public or in the big associations. It is the officials who increasingly become the pillars and beneficiaries of this system, not excluding the growing number of functionaries in the multiplying and spreading international organizations.

In this respect, then, we may ask whether the out-and-out welfare state does not counteract one of its own major purposes. The same question arises in another context. Like the welfare state's claim that it loosens class stratification, its claim to be an instrument of equality is very doubtful. While it certainly does work towards equality in the sense discussed so far, it does not do so in another, a crucial and wholly desirable sense. The continual compulsory redistribution of income undoubtedly furthers material equality. But at what price? This policy inevitably implies a growing concentration of power in the hands of the administration which directs the income flow, and this no less inevitably implies growing inequality in the distribution of power. Would anyone deny that the distribution of this non-material good, power, is incomparably more important than the distribution of material goods, since the former is decisive for men's freedom and unfreedom?

To say this is to say no less than that the modern welfare state, in the dimensions to which it has grown or threatens to grow, is most probably the principal form of the subjection of people to the state in the non-Communist world. The welfare state does not solve, or solves only partially, the problems which it is intended to solve; on the contrary, it makes them less susceptible to serious and genuine solutions. By contrast, it causes the power of the state to assume giant proportions "until each nation is reduced to nothing better than a flock of timid and industrious working animals, of which the government is the shepherd." It forces us to accept the idea that Tocqueville's vision has every chance of coming true now, after a hundred years.

171

The Problem of Social Security in a Free Society

Thus we have to be alert to the grave dangers which this development entails for the health of state, economy, and society alike and
for freedom, sense of responsibility, and naturalness in human
relations. The desire for security, while in itself natural and legitimate, can become an obsession which ultimately must be paid for
by the loss of freedom and human dignity—whether people realize
it or not. In the end, it is clear that whoever is prepared to pay this
price is left neither with freedom and dignity nor with security,
for there can be no security without freedom and protection from
arbitrary power. To this exorbitant price must be added another,
as we shall shortly see, namely, the steady diminution of the value
of money. Surely, every single one of us must then realize that
security is one of those things which recede further and further
away the more unrestrainedly and violently we desire it.

We can counter these dangers only if we refuse to drift with the
current. First of all we have to guard against confusing slogans.
One of the most dangerous and seductive of them is the expression
"freedom from want," which was coined by that master of the
alluring phrase, the late President Roosevelt, as part of the familiar
list of four freedoms.

We only have to think a little to realize that this is, in the first
place, a demagogic misuse of the word "freedom." Freedom from
want means no more than absence of something disagreeable,
rather like freedom from pain or whatever else may occur to us.
How can this be put on a par with genuine "freedom" as one of
the supreme moral concepts, the opposite of compulsion by others,
as it is meant in the phrases freedom of person, freedom of opinion, and other rights of liberty without which we cannot conceive
of truly ethical behavior and the acceptance of duties? A prisoner
enjoys complete "freedom from want," but he would rightly feel
taunted if we were to hold this up to him as true and enviable
freedom. We would do well to refuse to follow this ratcatcher's

172

tune of "freedom from want" right into a state which robs us of true freedom in the name of the false and where, unawares, we hardly differ from the prisoner, except that there might be no escape from our jail, the totalitarian or quasi-totalitarian state.

If we pursue this line of thought, we discover something rather strange. The truth is that what is meant by "freedom from want" is practically inseparable from compulsion, that is, the exact opposite of freedom. The reason is as follows.

To be in want means to be in a situation, for whatever reasons, in which we lack the means of subsistence and are unable to procure them by current earnings because we are ill or unemployed or bankrupt or too young or too old. We are freed from this want only if we can dispose of means from a source other than our current production. Thus provision must be made for us to be able to consume without producing at the same time.

The simplest and least problematic case is that we consume what we have set aside out of previous production. One important instance is to own a house, built or bought in better days, which will provide us with the vital good of shelter in bad days, too. But apart from that, the practice of accumulating goods against needy times is not the rule, either for the individual or for society as a whole. This is, in fact, not what happens in our highly differentiated society. If we have laid aside money and now use it up, this is not the same as if we ate up butter and lard previously produced and waiting for us in some store. Such stores would, on the contrary, be symptoms of grave disturbances in the circular flow of the economy. Normally, the consumption of our nest egg means that we are provided for out of current production by virtue of a title thereto acquired through earlier productive effort and certified by society in the form of money. In other words, if we think it through carefully, we live in times of need by consuming what someone else produces and does not himself consume. If we neglect for the moment certain qualifications and refinements to which we shall return later, this is what relief means in the context of society as a

173

whole: contemporary work also produces on behalf of those who, in straitened circumstances, consume without producing.

By what title the needy draw on the current flow of production is quite another question. The pursuit of this question leads us to a cross-roads where one arm of the signpost points to the welfare state.

Emergencies can be provided for either by the individual's own providence or by extraneous relief. It is self-providence if I have, by my own exertions and on my own responsibility, provided for the vicissitudes of life through saving or insurance; it is extraneous relief if I shift this burden onto others. Extraneous relief may be voluntary; I may, for example, borrow or accept charity or the help of my family or some other group, which, in return, counts upon me when another member is in need of help. For the rest, it is compulsory, and since this compulsion would not otherwise be necessary, it is considered as a burden imposed by the power of the state. This is rightly expressed in the very name "social charges," which indeed are, in practice, indistinguishable from the tax burden.

Now it is evident that the slogan "freedom from want" is not meant as an appeal for more self-providence, for saving and insurance. It was not understood in this domestic sense of good husbandry, either by Roosevelt or by the masses. What is implied is extraneous relief, not voluntary but compulsory, and on a large scale. But in that case all that "freedom from want" means is that some people consume without producing while others produce and are forced by the state to forgo consumption of some of their own production. That is the sober and elementary fact.

It justifies three conclusions. *First,* we see once more how thoughtless is the notion of a sort of fourth dimension, a cornucopia out of which the claim of any class for help in genuine or alleged need can be satisfied. It cannot be repeated too often that what is given to the one must be taken from the others, and whenever we say that the state is to help us, we are laying a claim to somebody else's money, his earnings or his savings.

This brings us to the *second* point. If it is true that the modern welfare state is nothing but a steadily spreading system of government-organized compulsory providence, it must obviously compete with the other forms by which a free society provides for itself: self-providence by saving and insurance and voluntary aid through family and group. The more the compulsory system spreads, the more it encroaches upon the area of self-providence and mutual aid. The capacity to provide for oneself and for members of one's family or community diminishes and, what is worse, so does the willingness to do so. Worst of all, it is only too evident that there can be no stopping on this road because the less able and willing the welfare state's citizens become to provide for themselves and help others, the more pressing becomes the demand for further expansion of public mass providence, leading to further curtailment of the ability and willingness to provide for oneself and voluntarily help others. It is yet another vicious circle.

This constitutes a further urgent warning that we must not allow the welfare state to develop to its critical point. Should, unfortunately, this point already be reached, then we must do everything in our power to bring about a contraction of this disproportionate welfare state and to widen the area of self-providence and voluntary aid in spite of strong political and social resistance. To widen this area is one of the foremost tasks today if we want a sound and well-balanced society. This surely needs to be stressed no more; we are at the cross-roads of a free and a pre-collectivist society.

The road we must take is plainly mapped: not more welfare state but less; not less self-providence and voluntary aid but more. Here I come to my *third* point. We cannot, nowadays, do without a certain minimum of compulsory state institutions for social security. Public old-age pensions, health insurance, accident insurance, widows' benefits, unemployment relief—there must naturally be room for all these in our concept of a sound social security system in a free society, however little enthusiasm we may feel for them. It is not their principle which is in question, but their extent, organization, and spirit.

175

The extent, organization, and spirit of that minimum of compulsory public providence will be mainly determined by the purpose in view. This is where opinions finally divide. It is a matter of the personal approach *versus* the collectivist, freedom *versus* concentration of power, decentrism *versus* centrism, spontaneity *versus* organization, human judgment *versus* social technique, responsible husbandry *versus* irresponsible mass man. After everything we have said, we surely need neither specify nor justify our choice. The purpose of minimum compulsory public providence must not be abused to set up a general system for taking care of all citizens and an all-pervasive social security organization. Least of all must the problem of relief for the weak and helpless be taken as a pretext for leveling out all differences in income and wealth. We need not repeat where that road leads. It is the road of social revolution, with all of its far-reaching consequences.

If we reject all this, our purpose can only be to support the really weak and helpless, to give them enough sustenance so that they do not become destitute—no more and no less. This assistance should be subsidiary only, to help out where the individual's own resources or voluntary aid prove inadequate; it should not become the normal form of satisfying the need for security.

The proper measure is not transgressed as long as such public providence does not weaken the impulse towards voluntary self-help and group aid to supplement the bare subsistence minimum. The experience of Switzerland and the United States shows that, the introduction of comprehensive obligatory old-age insurance notwithstanding, total savings and private life assurances have risen considerably. This proves that such a desirable development is possible, whereas Great Britain and the Scandinavian countries, the models of the extreme welfare state, furnish discouraging examples of the contrary.[15]

These considerations surely make one thing crystal clear. It is that the problem of social security in a free society is not primarily one of the technique of social insurance or social administration, and still less one of political expediency, but one of social philosophy. Before we go into actuarial mathematics, we must have a clear picture of what we mean by a sound society. Only then shall we know on which side to put the emphasis: whether we are to strengthen the individual's resources, sense of responsibility, and thrift, as well as the natural solidarity of small groups, above all the family, or whether we are to give yet further impetus to the already almost irresistible modern tendency toward collectivism, the omnipotence of the state, mechanical organization, and the tutelage of man. Once we consider the direction in which we move in the two cases, it must be evident that ultimately the choice lies between the individual and family on the one side and collectivism on the other or, to put it plainly, between the climate of freedom and its opposite. To consider this a mere empty phrase is to fail to grasp what is at stake today.

It would be frivolous to ignore these considerations. They are necessary if we want to know in what direction we are moving as we make decisions on particular questions of social policy technique. We may feel that we cannot avoid many a step in the wrong direction, but at least we should take them reluctantly, knowing that we are accepting a necessary evil and that each additional step along this road increases the danger. We certainly should take no such steps without a very clear and firm idea of what is the rule and what the exception, what the sound norm and what the possibly tolerable deviation. If we are earnestly concerned about the ultimate foundations of our civilization, our rule and norm and our cheerfully accepted ideal should be security through individual effort and responsibility, supplemented by mutual aid. It is the ideal of the "well-ordered house," and we cannot abandon it without shaking the very foundations of a free society and making its difference from Communism no more than a matter of degree.

In no circumstances should we allow ourselves to be misled by the argument that it is no longer possible in our age to give first place to self-providence and voluntary mutual aid and to reduce public providence to a subsidiary minimum. This is defeatism, and it does not become any more convincing for generally being associated with a barely concealed distaste for the former course of action. It belongs to the category of insincere resignation, which, by capitulating before allegedly immutable facts, contributes to its own justification. If we start out from the argument that in our age the problem of social security for the masses can be solved only by means of collective compulsory action and that any considerable widening of the private zone is an illusion, then we shall end up by loading so much onto the compulsory system that the masses, burdened as they are with correspondingly high contributions and taxes and relieved of any care for their future, will be neither able nor willing to provide for themselves. If only the compulsory system is sufficiently thoroughgoing and comprehensive, it is easy triumphantly to denounce self-providence as a pipe dream. But all that is thereby proved is the familiar fact that the welfare state has a fatal tendency to get into a vicious circle from which we must escape.

It would be somewhat surprising if no one had thought of declaring self-providence by the masses as not only hopeless but as a catastrophe for the economy. It has indeed been said that our modern economic system could not possibly digest so much saving. If "excess saving" is not to plunge the economy into deflation, depression, and unemployment, the capital thus accumulated must be absorbed by investment. But where are the investment opportunities on the scale presumed? Our reply is that this is a pseudo-Keynesian exaggeration and oversimplification. It is a pity that we do not know what Keynes himself would have said, as chairman of an insurance company, about this attempt at playing off his theory against people's endeavor to better themselves by saving and insurance.

In the first place, this argument neglects the fact that if the practice of self-providence is to be at all possible, it presupposes high average incomes resting on high national productivity. This, in turn, presupposes genuine economic growth not artificially whipped up by inflation, and such growth depends upon appropriately large investment, which, if inflation is to be avoided, must be covered by true saving. If, then, growing self-providence leads to a higher savings ratio, we have an urgent need for these extra savings in order to avoid a situation in which the high mass incomes, which we must assume by definition, do not rest upon the precarious foundation of inflationary investment. Moreover, as I have mentioned before, a by no means negligible part of self-providence takes place in an area where the question of balance between saving and investment does not arise at all, namely, when people acquire property in the form of a house and garden, which is one of the most important and desirable forms of self-providence. It is surely absurd in this case to speak of balancing saving and investment.

To take a concrete example, a country like Germany would gain a great deal if only it attained to the degree of self-providence common in Switzerland or the United States. But in both cases the problem is one of inflation, not deflation. Although Switzerland is the classical country of savers, policyholders, and private pension funds, saving is insufficient to contain inflationary tendencies and to finance investment projects. The unusually high amount of saving in Switzerland has not given rise to any problem of how to prevent investment from lagging behind saving and thereby releasing deflationary tendencies. Thus the problem itself is a pseudo-problem which may drive us into the fold of a compulsory state system of social security. To the extent that saving has risen in Switzerland and in the United States with the growth of the economy, this increased saving at the same time is the condition of further non-inflationary growth.[16]

Another argument is sometimes adduced. It is that a compulsory state system of social security has a great advantage over self-providence in that it does not require prior capital accumulation

179

and needs to raise only the currently necessary means each year, thereby living from hand to mouth. Is this not much cheaper, it is said, and does it not, therefore, make possible much more comprehensive and generous social benefits for the masses?

This simple procedure, which has been called the pay-as-you-go method, is in fact applicable also to mutual aid in smaller groups, but on the large scale necessary for mass social services, it is obviously reserved to the state, with its powers of compulsion. However, this is anything but an advantage. It is not enough to appeal to the elementary principle that any social payments must, in reality, always be covered by current production. Another circumstance needs to be stressed, and it adds an important qualification to an axiom we have mentioned before: the extent of current production is decisively influenced by previous investment, and unless this investment is to have inflationary effects, it must as a general rule be covered by saving.

Therefore, a pension system resting on capital accumulation makes a considerable contribution to national capital formation as a determining factor of the national product. The system thus tends to increase the national fund of goods out of which the social payments, translated into goods, are made. A pay-as-you-go system, on the other hand, would stop up this source of capital formation and, unless a substitute can be found, hamper the growth of the social product. The more comprehensive such a system is, however, the less can a substitute be counted upon. It is bad enough that social-insurance funds may impair the genuine and traditional forms of self-providence and thereby private capital formation, but at least they fill the gap with a sort of collective saving. But a pay-as-you-go system would retain the disadvantage of weakening self-providence without compensating the diminution of private capital formation by collective saving.

It is difficult to think of a worse combination. But in a mass democracy there is an extraordinary temptation to choose this method because it offers the possibility of organizing a compre-

180

hensive and generous system of social payments without the awkward restriction of capital coverage. The temptation is all the greater since it is possible with this system also to adjust social payments currently to the price and wage increases that are due to chronic inflation. German politicians recently succumbed to this temptation in spite of warnings to the contrary. This is an alarming sign of the times. There is every reason to expect that this example will soon be followed elsewhere.[17]

However, considerations of this kind, though necessary and not to be evaded just because of their technicalities, have the dangerous tendency to obscure from view the wider questions which should never be lost sight of in any discussion of the development of the modern welfare state. It may be well to return to these questions once more, if only to conclude the argument by stressing them. Two ideas above all need to be made quite clear.

The first is one with which we are already familiar. It is impossible fully to grasp what is at stake today without constantly keeping in mind that the system which goes by the name of welfare state is in the process of altering our society in one particular direction: in the name of equality and mediocrity it is choking everything above the average. We are moving towards a situation where the "common man" is relieved of his responsibilities and the "uncommon man" robbed of his keenness. But since above-average ability is the real condition of production, and at the same time so rare that it needs most careful cultivation and encouragement, it is not difficult to see what fate we are thus preparing for ourselves. The prospect is made even darker by the fact that the masters of the Communist empire are shrewd enough to encourage and reward above-average ability as best they can. What Charles Morgan wrote some years ago has lost none of its pertinence today: "The central crime against a society impoverished as ours has been is by no means to be happier or abler or healthier or more enterprising than other men, but to be a dependent mediocrity, fattening upon the state."[18]

The other idea may be expressed by a simple image. Let us imagine that we are looking at one of the greatest works of Western art, Tintoretto's wall and ceiling paintings in the halls of the Scuola di San Rocco at Venice. This was one of those benevolent fraternities which in their time and in their fashion solved the problem of helping the poor and without which the city of the lagoons would hardly have survived a thousand years without revolution. The devotion of the friars was matched by the artist's, who is said to have taken no fee for his enormous work.

Now suppose, for the sake of argument, that there were a painter today of Tintoretto's stature. Can we imagine any welfare-state authority asking him to decorate its offices? And can we imagine a Tintoretto, absorbed in his task, painting his great work in selfless devotion to the glory of God, beauty, and love of man?

These are cruel questions. But then *we* have the modern welfare state.

The Welfare State on the International Plane

Be that as it may, the fact remains that whatever we may have to say against the welfare state, the problems which it tries to solve are real enough. There are the economically weak, who are to be helped by the economically strong; there are the poor and the rich, between whom there ought to be no yawning gulf. If this is true of individuals, why not of whole nations? Are there not "poor" nations and "rich," "economically underprivileged" and "privileged" ones, and can the discrepancy between them not furnish plausible enough grounds for a claim to "equalization" such as underlies the welfare state? Why, then, not have a welfare state on the international scale, with some nations giving, either voluntarily or compulsorily, and others receiving?

The idea is tempting and as such is by no means new. We met it twenty or thirty years ago in the Fascist and Nazi catchwords of the haves and the have-nots. We remember the violence with which

Mussolini launched what he called the class struggle of the proletarian peoples against the satisfied and possessing peoples, and the Nazis demanded living space for themselves.[19] At that time it was advanced industrial nations which insisted upon their right to a fair share in the raw-material resources and colonization areas of the underdeveloped countries; now it is the latter themselves which have taken up the battle cry of international "social justice." In these countries and in the Western world the economic development of underdeveloped areas has become a slogan pressed home with more force than most others which we hear today, and there is no mistaking the strong emotional overtones which are so familiar a feature of the national welfare state.[20]

The claim to overtake the lead in wealth established by others and the desire for equalization in the well-being of whole nations are pitched in the same key as the claims of the "underprivileged" against the "privileged" which, on the national plane, have led to the conception and creation of the welfare state. No perceptive observer of the current discussion on the development of underdeveloped countries can help being struck with the tone of demand, defiance, and grievance on the part of those who believe themselves to be "disinherited," with the note of envy on their side and of "social conscience" and fear of envy and resentment (and their exploitation by Communism) on the other. It is hardly surprising in these circumstances that in the Western world the program of the economic development of underdeveloped areas finds its most active supporters among those who, in their own countries, advocate the welfare state, economic planning, socialization, and inflationary policies. The development of underdeveloped countries has become one of the most important fields for the champions of these ideologies.[21]

Our first reply is that the analogy does not hold. One cannot equate nations and individuals without becoming guilty of that very common sophism which A. N. Whitehead calls the "fallacy of misplaced concreteness" or political anthropomorphism.[22] More-

over, there is no question here of provision for the future and social security against risks; what is involved is a claim of those who
are economically less successful to the wealth of those who are
more successful. The champions of this idea have in mind not the
legitimate but the unlawful and revolutionary aspects of the slogan
of the welfare state, but they are not honest enough to say so
clearly. Nor do they recognize that such international equalization
of wealth could be brought about in no other way than by coercion
on the part of an international state. If they gave this some thought,
they would have to admit that it would be utopian to count on the
creation of world government. But our weightiest objection is perhaps that they misunderstand the problems facing the underdeveloped countries.

What is the meaning of economic development in an underdeveloped country? The definition is not an easy one. We may perhaps best put it this way: such countries try to repeat the process
of economic growth which Great Britain was the first to achieve
at the time of the industrial revolution and which has since taken
place in one country after another. We are now beginning to have
a better insight than before into the nature of this process. Above
all, we are beginning to realize how difficult the start is bound to
be and how many sacrifices it entails and how great is the diversity
of the conditions which determine the pace and success of development. At the time of early capitalism in England, and later in the
other industrial countries of Europe, the United States, Canada,
and all other countries, including Russia, the principal questions
were these: Where is the necessary capital to be found? Where are
the workers to be found whom industry needs? Where is the entrepreneurial spirit to come from, with its initiative and industrial
leadership, which are indispensable for the take-off into the dynamic
forms of the modern industrial economy? Where is technical experience to come from which clearly is equally indispensable?
And finally, where are the agricultural surpluses to be found with
which to feed the growing industrial and urban population?

184

The pioneer country of the modern industrial economy, Great Britain, had the most difficult time in all these respects because it had to rely on its own resources. This is particularly true of the problem with which the first of our questions is concerned and which is of paramount importance, namely, the problem of capital accumulation to set economic development in motion. The task was no less than to make the "critical start" of economic development without any appreciable influx of capital from abroad and to draw on the country's own resources in accumulating the capital necessary for the construction of machines and factories, railways and ports, and other investment. This could only be done at the cost of restricting the consumption of a people which, at that time, was still poor and at the cost of other burdens, hardships, and sacrifices. The price of the "critical start"—of the necessary sharp kink which sends the capital-supply curve steeply upwards at a stage when the fruits of development in the form of growing social product have not yet matured—was, in England, a situation which we have become accustomed to describing as the misery of early capitalism, a habit due to the influence of Marxist propaganda and of a bias in the tradition of theoretical economics which has only now been overcome.

But there is no need to go to the other extreme. Whatever light modern economic historians may have shed on this period, it was grim enough.[23] Nevertheless, a sobering thought imposes itself. Nowadays, when this process of "autarkic industrialization" is repeated in many of the underdeveloped countries, we understand better than before that a period of at least relative restriction of mass consumption for the purpose of a rapid increase in domestic capital formation is an indispensable condition for the economic development of a country which cannot count on foreign capital aid. In England, capitalism had, as it were, to starve itself upwards, and it is not surprising that the British industrial revolution was not at once associated with that growth of mass incomes which was expected of the new technical miracles. What is astonishing is only how quickly even the pioneer country of industrialization managed, in spite of enormous difficulties, to get beyond the "critical start," to raise the situation of the masses steadily, and to improve the originally so oppressive labor conditions. All this was done without the frightful and prolonged privations of Russian Communism, without forced-labor camps, secret police, and firing squads.

The mention of Russian Communism brings us to the essential point. However difficult the beginning may have been in England because the central problem of capital supply had to be solved by the country's own efforts, the nations which followed England's example had far fewer difficulties. I have in mind the development of Western and Central Europe, the United States, Australia, South Africa, Argentina, and Canada. Their economic development was facilitated by their being able to draw on the capital accumulation and on the economic and technical experiences of the earliest industrial countries, England and those immediately following—provided only that the other conditions of economic development indicated in our list of principal questions were fulfilled. This was, in fact, mostly the case.

All this happened quietly, without anybody making a big "problem" out of it, without international organizations, programs, conferences, committees, and officials, without any pangs of moral and political conscience on the part of the developed countries and without fear of the possible consequences of inadequate help, and also without that mixture of begging, threats, and blackmail with which the underdeveloped countries appeal to conscience and fear. Nobody dreamed of assuming compassionate attitudes towards the poor devils of the American prairies, the Australian bush, or the pampas of the Argentine; trusting their ability and good faith, one lent them money at 5 per cent and considered that this was a good bargain for both sides. But one thing was, of course, always taken for granted and therefore not even discussed, namely, the existence

186

of the whole body of conditions and institutions which justified such trust and formed a free and firm bond between the developed and underdeveloped countries: freedom of international movement of goods, capital, people, and ideas, the rule of law, the market economy, respect for money and everything this implies.

Russian Communism was the first great example of the fact that a developing country which chooses an economic and social system incompatible with the conditions of such a free flow of aid from the developed countries makes such aid impossible. If such a country insists on economic development, it condemns itself to the extraordinarily difficult path of autarkic industrialization after the British pattern. The Kremlin masters thought for a while that they could escape this inexorable logic, but Stalin's first Five Year Plan showed that they had grasped it. Communism has meant not only the hard necessity to make the critical start by the privations of Russia's consumers and peasants, but also the choice of a method, the collectivist, under which the hardships of such national "self-financing" of development are inordinately aggravated by the inherent weaknesses of a collectivist economic order.

What is the misery of early British capitalism in comparison with the immense sacrifices of the Soviet experiment? The British had to wait a little while for the increase in mass prosperity and an improvement in labor conditions—but what is this in comparison with the long and still continuing sufferings of the masses in the Communist state? Nor should we forget that Moscow's autarkic and collectivist method made the solution of another development problem much harder, namely, the problem of feeding the growing industrial and urban population. In England and in the other Western countries, development was accompanied by a steady and considerable increase in agricultural yields, and at the same time, the free world economy enabled the produce of the vast new cultivated areas of the New World to be used for feeding the industrial countries; but in Soviet Russia, Communist economic methods led

187

to a decline in agriculture which even now does not seem to have been made good, if we are to judge by Russian statistics and the observations of Moscow's rulers.

Today's underdeveloped countries have to decide for themselves whether to solve the key problem of capital supply according to the West's international and market method or according to Moscow's autarkic and collectivist one. With the first method, the problem will be solved in the way which was normal and natural hitherto, and still is in the case of a country like Canada, that is, by a free and spontaneous influx of foreign capital. But if these countries' own policies of nationalism and socialism destroy the conditions of such capital supply, they have no right to complain of its absence, let alone to claim international charity.

Such underdeveloped countries thereby maneuver themselves into a position where they clamor all the more vociferously and urgently for the kind of capital aid which we might call political and which corresponds to the concept of an international welfare state. If individuals in the Western countries have insufficient confidence in the government of an underdeveloped country to entrust it, of their own free will, with their savings, then these individuals are to give up their savings by compulsion, via their own governments, without reward and without hope of repayment—spurred and applauded by international officials who, themselves, pay no taxes.

If an underdeveloped country cannot tap the source of foreign capital markets because of its own nationalist and socialist policies, then it must seek a supply of political capital. The money which will not flow freely has to be pumped up by means of diplomatic conferences, propaganda, and open or disguised threats, even at the risk that the flow may evaporate or trickle away in the heat of the same passions which have already dried up the original source. When it becomes impossible to turn to the market and private investors, the governments of the West, and through them the taxpayers, have to be mobilized.

This is the simple state of affairs to which the confusing multiplicity of action in this field can be reduced. Many underdeveloped countries refuse to satisfy the conditions necessary for a voluntary flow of capital from the West. They reserve to themselves all sorts of rights and devices, such as taxation, expropriation, exchange control, expulsion of foreign technicians, company law discrimination, and so on, and they refuse to pay in interest, dividends, and salaries the price without which no capital aid can be offered even in the most favorable case. All the more passionately do these countries proclaim their right to receive such aid for nothing and by way of the compulsion which Western governments have to exert upon their taxpayers in order to raise the required capital. By the same crazy logic, the sums demanded become more and more fantastic.

In these circumstances it is more than ever necessary to stress the sober facts which cut the ground from under this concept of an international welfare state. Those underdeveloped countries which, by their policies and principles in economic and social matters, create the necessary conditions—the right climate—for private investment, obtain Western capital through the market. Fortunately, this species has not yet quite died out. The others, which do not create these conditions, have no right to complain about the consequences. As they make their own bed, so must they lie on it. If a country resorts to political means to obtain capital aid by begging, defiance, or threats, it cannot invoke the argument of necessity. If it sets its policies by the lodestar of nationalism and socialism and persists in doing so, it must pay the price. If it does not want to pay the price, it must alter its policies.

That is the clear alternative. We should not let it be obscured any longer, not even by reference to certain undeniable facts which distinguish many of today's underdeveloped countries from the normal cases of the past. It is true that some of the most important of the underdeveloped countries suffer from unprecedented overpopulation, but this is no reason why they should, in addition, pursue policies which scare foreign capital away and so make the position worse. Egypt has become a warning example of this kind of thing. It is also true that, unlike the classical Western cases, many under-developed countries lack some of the essential conditions of industrialization, especially potential entrepreneurs and skilled labor; but in such cases it may well be asked whether it would not be better to do without industrialization rather than to enforce it by the methods of nationalism and socialism.

The Theoretical Background of Chronic Inflation

After this excursion, let us return to our main subject. We now have to examine the close link which exists between the modern welfare state and the chronic inflation of our age, which we have had occasion to mention more than once. In order to see the connection, we must examine and analyze chronic inflation itself.

We need not waste much time on refuting attempts to deny that inflation has become chronic. It does, of course, afflict different countries in different degrees, according to the resistance, more or less energetic, of the government and the central bank. It occurs throughout a whole range of temperatures: "hot" inflation in countries where the value of money is melting away rapidly, unless "open" inflation is transformed into "repressed" inflation[24]—but this is a measure which only Communist countries can apply at all successfully today; "temperate" inflation in most underdeveloped countries and pronounced welfare states; and "cold" inflation in countries which, like Switzerland, West Germany, and Belgium, have hitherto been most successful in combating inflation. These international differences in the temperature of inflation generate difficult problems for international payments, of which we shall have more to say later. But inflation in some form or other is today endemic in all countries, and no one denies it, not even those who might have an ideological or practical interest in pretending that there is no inflation.

190

By contrast, it is not at all redundant to stress the point that to-day's world inflation has not suddenly overtaken us during the last few years but is part of a long-term inflation. Apart from short and insignificant interruptions, there has, since 1939, been a rising tide of increasing volume and decreasing purchasing power of money—so much so that one could rightly speak of the "great inflation."[25] The dollar, the Swiss franc, not to mention sterling and other currencies ravaged by the war, have during the last twenty years lost half, or more, of their value, in terms of increased prices. The end of the process is not in sight. Throughout history, inflation has often afflicted one country or another. But during the last five hundred years, it has happened only four times that the deterioration of money was not limited to one nation but spread to the whole economically developed world: the era of the Spanish silver fleets, the French Revolution and Napoleonic Wars, during and after the First World War, and now, in our own time. All of these were catastrophes, but it looks as if the last were the worst. It differs from its predecessors in some essential characteristics to the point of being historically unique.

First of all, repressed inflation had never before been experienced; this particular form of inflation was unknown in the past because it presupposes a hitherto unknown degree of state power. It was for our time that this combination of inflation and collectivism was reserved. In other words, this is the first great inflation of the collectivist age.

This leads us to the *second* novelty. Our inflation is the first to be marked, unequivocally and almost exclusively, by the ideologies, forces, and desires of modern mass democracy. It is a democratic and social inflation and comes close to meeting the prediction which a distinguished American social philosopher made more than thirty years ago: "It is not yet clear that it is going to be possible to combine universal suffrage with the degree of safety for the institution of property that genuine justice and genuine civilization both require. . . . If property stands for work in some sense

191

or other of the word, and if money is the conventional symbol of property, the ends of justice tend to be subverted if this symbol fluctuates wildly; thrift and foresight become meaningless; no man can be sure that he will receive according to his works. Inflation of the currency amounts in practice to an odious form of confiscation."[26]

The *third* essential feature peculiar to our own "great inflation" is that its time limits cannot be strictly defined and that its period does not end with a definite return to monetary stability. Rather, it is the acute stage of a chronic pathological process fed by forces which are now permanently operative, and as such, it is not susceptible to any quick or lasting cure. The inflation of our time is intimately connected with some of its most obdurate ideas, forces, postulates, and institutions and can be overcome only by influencing these profound causes and conditions. It is not just a disorder of the monetary system which can be left to financial experts to redress, it is a moral disease, a disorder of society. This inflation, too, belongs to the things which can be understood and remedied only in the area beyond supply and demand.

We have seen earlier that the inflation of our time is a bitter irony of history. It has come to pass in contradiction of gloomy forecasts and of the economic policy concepts deriving from them. These concepts were based on the fear of deflation and it was an important cause of the "great inflation" that they were but very slowly recognized and admitted as inapplicable. The blame for inflation must be laid at the door of the whole trend of postwar economic policy in most countries, that mixture of planning, welfare state, cheap-money policy, fiscal socialism, and full-employment policy; but to understand this policy trend, we must go back to the revolution in economic theory which furnished the ideas and catchwords of inflationary policy and which is, above all, linked with the name of J. M. Keynes.

His theory is seductively brilliant and elegant. However, this is not the place to discuss the separate links in the chain of thought

192

which led Keynes and his followers to their bold conclusions or to show why the chain does not hold.[27] Our main concern here is the result of this "new economics" for the theoretical basis of economic policy. We may say briefly that a whole generation of economists was so exclusively brought up to operate with economic aggregates that it forgot the things which until then were the real content of economic theory and which should never be forgotten: namely, that the economic order is a system of moving, and moved, separate prices, wages, interests, and other magnitudes. Keynes's aggregative functions made the plain mechanism of single prices look outdated and uninteresting, and we witnessed the development of a sort of economic engineering, with a proliferation of mathematical equations. In the past, to be a good economist meant being able to assess the relationship between currently operative forces, and sound judgment, experience, and common sense counted for more than formal skill in handling methods illegitimately transferred from the natural to the social sciences; but the limelight came increasingly to be occupied by a type of economist who knew how to express hypothetical statements about functional relationships in mathematical formulas or curves.

This new method was one part of the training of the new generation of economists and economic policy makers; another was the idea that saving is, at best, unnecessary and may be harmful. It follows that a policy measure is good when it increases effective demand and bad when it threatens to diminish effective demand. But if saving and good husbandry are represented as enemies of economic progress, the leveling of income differences, which socialists had so far demanded only on moral pretexts, can be put forward as a command of economic reason. The danger of inflation was reduced to a vague and remote possibility; the thing to be feared constantly was what was usually somewhat imprecisely described as deflation. Budgetary deficits, leveling taxes which diminish both the ability and the willingness to save, artificially low interest rates, a combination of growing popular consumption

193

and investment stimulation, expenditure and credits on all sides, mercantilist foreign-trade policies with the twin purposes of mitigating the effects of those other policies on the balance of trade and of creating export surpluses as a further stimulant for the domestic money flow—all of these practices now received the blessing of economic science. Any protest was dismissed as stupid and old-fashioned, and when a government such as that of postwar Germany resolutely adopted an opposite course, it had to face a drumfire of criticism.

Now, when no one can deny any longer that inflation is indeed with us, we ought not to forget that this is the seed which Keynes has sown. No honest person can overlook how abundantly it is bearing fruit. The profound economic and social disorder which faces us in this inflation was prepared by an intellectual one. Without Keynes, or, rather, without *The General Theory of Employment, Interest and Money*, the science of economics would no doubt be poorer, but the nations would be richer to the extent that the soundness of their economy and currency would be less impaired by inflation. In so far as the full danger of inflation is now generally recognized, there may be wider understanding for the reasons why the writer and his friends, who quickly took a decided stand against the destructive effects of deflation in the thirties, equally quickly took a decided stand against the Keynesian doctrines later, when the danger of inflation became apparent.

There can be no doubt that Keynesianism contributed decisively to the utterly wrong postwar orientation of the Western world, which, taught only to fear and combat deflation, followed the banner of "full employment" right into permanent inflation. In spite of all the warnings of the old-style economists, the danger was recognized too late. It has become exceedingly difficult to face about towards the true enemy, inflation.

We should never have forgotten that over the course of centuries there has always been more danger of inflation than of deflation. Inflation is always a lurking temptation and at all times the way of

194

least political and social resistance. Both inflation and deflation are monetary diseases, but, unlike deflation, inflation has an initial pleasant stage for wide circles of the population, and above all for the politically most influential, because it begins with the euphoria of increased economic activity and other boom symptoms. By and large, things happen just as they are described in the second part of *Faust* in the famous paper-money scene: "You can't imagine how it pleased the people." But that is precisely the dangerous seduction of inflation: it begins with the sweet drops and ends with the bitter, whereas deflation is most disagreeable from the very outset and is marked by the unpleasant symptoms of depression, unemployment, a wave of bankruptcies, the closing down of factories, losses all along the line, and contraction of economic activity. It follows that of the two diseases, inflation is the rule and deflation the exception. In the course of the centuries, no wager has been more of a certainty than that a piece of gold, inaccessible to the inflationary policies of governments, would keep its purchasing power better than a bank note.

There was never much likelihood that governments would abuse their power to create money for the purpose of deflation, and today, in the age of paper currencies and the prevalence of inflationary ideologies and interests, this probability has, to all intents and purposes, become nil. All the greater is the danger that governments, swayed by weakness, ignorance, and lack of responsibility, may yield to these ideologies and interests and pursue policies which either cause, or at least favor or fail to obstruct, inflation. It is no exaggeration to say that hardly any government ever possesses absolute power over money without misusing it for inflation, and in our age of mass democracy the probability of such misuse is greater than ever before.

To wrest this power from governments and to make the monetary system independent of their arbitrariness, ignorance, or weakness was one of the essential functions of the gold standard; having so withdrawn money from politics, the other and equally im-

195

portant function of the gold standard was to create a truly international currency system.[28] Never has it been more essential to keep money out of politics than in our age of mass democracy. After the demise of the gold standard, a last counterweight against absolute government power over money remained, in the form of a certain degree of independence on the part of central banks. But this dam, too, has now broken in many countries, and in others it is so undermined that it is becoming less and less effective. Independent central banks seem to be among the *Bastilles* which give our modern Jacobinism no peace until they are razed to the ground.

Inflation is as old as the power of government over money, and equally old are the theories and ideologies which, while not always justifying inflation, at least excuse it. What is new in our times is that such theories were never before so daring and subtle and that the ideologies supported by these theories were never before so powerful as since the beginning of our own period of inflation. This is the background which we have to keep in mind in examining more closely the chronic inflation of the present day.

The Nature of Chronic Inflation

Today's chronic inflation is all the more uncanny because its nature is difficult to assess. It does not fit into any conceptual cliché but is something new in economic history. It therefore has to be traced to hitherto unknown causes. No wonder, then, that there is a great confusion and no end to guessing. We do not see a flood of money swelling visibly or any printing presses working day and night. Except in the simple case of a country where inflation is, in old-fashioned manner, due to budgetary deficits, most people look in vain for the source of inflationary pressure and do not know whom or what to hold responsible for it. Only one thing is plain: slowly but inexorably, everything becomes more and more expensive, and it is hard to see how this could change. In many countries there is no sign of a budgetary deficit, which is the customary source of inflation; in countries like Germany and Switzerland, even the familiar symptoms of balance-of-payments deficit and pressure on the exchange rate are absent. Earlier periods of long-term price rises, such as the so-called prosperity period before 1914, for various reasons do not lend themselves to a comparison— if only because they used to alternate with equally long depression periods, whereas we now know only too well that, apart from a comparatively short and mild recession as in the winter of 1957-58, such alternation has become highly improbable. Moreover, such recessions, with their measure of unemployment, do not necessarily interrupt inflation, as the example of the United States shows.

This is embarrassing. Some tried until recently simply to deny the process and to maintain that what was called inflation was just a bogy. It was quite normal, and always had been so, that prices rose slowly, some said, without stopping to think whether it really had always been so and whether, secondly, our era's rising productivity should not properly cause prices to fall, so that we should really begin to worry when prices failed to show any tendency to fall. Is it not already inflation if prices simply remain stable? Others abandon altogether the search for cause and responsibility and lay the blame on the broad shoulders of historical necessity— they speak of the "age of inflation" as if this were inevitable. Or else they vaguely blame institutions and circumstances about which nothing can be done. Often enough there is a clear implication that in any case it does not matter too much. We have got used to such oracular pronouncements as "institutional inflation" or "cost inflation." Or, finally, everybody accuses everybody else: the workers accuse the employers, the employers the workers, and the government accuses both.

It is indeed a nagging problem which is obviously not easy to solve. But one thing should be clear in this case, too, and may therefore serve as a first point of orientation in any discussion. Whatever the nature of the process of chronic inflation, there must be an excess of total demand over total supply. Since the supply

197

of goods is steadily rising, the imbalance between supply and demand is obviously not due to a sudden deficiency of supply, and excess demand must therefore be due to additional money reaching the market. Neglecting the possibility that this additional money may have been previously hoarded and is now put into circulation or that it is left idle for shorter periods and more quickly spent (increase in the velocity of circulation of money), the extra money can come only from where money is created. This means that it can ultimately be traced back to the central bank, which not only creates cash but in most countries disposes of means to influence bank liquidity and so to favor or disfavor the creation of bank money. It is in the central bank that we have to look for the tap which needs only to be closed firmly to stop the dripping. This fact can hardly be denied, and it puts the ultimate responsibility squarely on the central bank. It is here that all the tangled plumbing converges.

Theoretically, at any rate, it is indisputable that the central bank could use the instruments of credit restriction to reduce the quantity of money sufficiently to counteract all inflationary tendencies, whatever their origin. In practice, on the other hand, we are immediately faced with the difficulties connected with the very nature of contemporary inflation tendencies. The next step in our analysis must therefore be to examine the sources of today's inflationary pressure.

There are in all four such sources of inflation. The first two exist only in a few countries, whereas the last two operate everywhere and constitute the real problem of present-day chronic inflation.

The first source is *fiscal inflation*. In its simplest form it is due to a budgetary deficit. This needs no further explanation, but, as we have mentioned before, this sort of classical inflation is today serious only in a few countries, mainly in Brazil and, until quite recently, in France. More complicated forms of fiscal inflation are at work in a far greater number of countries where the budget is used to give demand a shot in the arm; this may take the form, as in Germany, of spending previously accumulated budget surpluses, in which case the effects are much the same as if inflation were due to budget deficits, or it may take the form of transforming savings into consumption through the budget. We shall have more to say about this later.

The second source is *imported inflation,* as it has been called. What is meant is an inflationary impulse from abroad that comes in such a manner that another country's inflation causes an influx of currency which the central bank transforms into domestic currency. In the absence of forces compensating the inflationary effects of the influx of foreign currency, foreign inflation can, in fact, be propagated to a country which is more successful than others in containing its own domestic inflation. This can happen on two conditions: first, the difference between inflationary pressure in the two countries concerned must be large; and second, exchange rates must remain unaltered in spite of this divergence of national currency policies. This gives us the key to overcoming this form of inflation; it can be brought to an end by modifying either of the two conditions. In any event, this form of inflation is today significant only in Germany and raises least problems.[29] We need not dwell on it any further in this context, though it may become a major problem for many Western European countries in the event that the inflation in the United States should prove unmanageable.

The third source, *investment inflation,* takes us right to the center of the problem which today preoccupies and alarms all Western countries. It is an excessive strain on economic resources, a symptom of the boom finding expression in overinvestment. The position becomes critical when the use of productive resources by investment and the immediate creation of incomes due to the construction of plant and machines are not matched by a corresponding immobilization of purchasing power through saving and the reserves of unutilized resources are exhausted. An inflationary boom thus occurs when current saving, which represents private consumption forgone, is no longer sufficient to balance the increase in demand due to investment and when investment is financed by a substitute for saving, that is, credit expansion. To the extent that this happens, investment presses on the dwindling reserves of unutilized resources and gives rise to inflationary excess demand.

One point needs to be stressed here. It is a matter of excess investment in relation to the true saving of the population and not from the point of view of the single business firm where investment may be justified technically and by market conditions. The single entrepreneur will always, and sometimes indignantly, tend to defend his investment program against the reproach that it may be excessive. We have another kind of excess in mind, an excess, that is, over the economy's readiness, expressed in terms of saving, to release the resources necessary for the construction of factories or power stations as well as the consumer goods corresponding to the wages paid therefor. Such an excess of investment over saving constitutes additional demand not covered by goods; it overstrains the economy, and, as always, the economy responds with inflation. The single entrepreneur can hardly be blamed if, under the impact of market conditions and competition, he takes advantage of possibilities to finance his investment program, even at the risk of thereby contributing to the national overstrain. It is not up to him to avoid this; it is up to those who are responsible for monetary and credit policy. He should, however, willingly support their efforts to re-establish equilibrium between saving and investment by tightening credit conditions. It is high time, too, that the state, which is subject to neither market nor competition, should curtail its own investment programs.

Now if today's inflationary pressure is in part due to excess investment in relation to saving, we may ask what are the reasons which have pushed up investment so steeply in our times. In reply, we may point to impetuous technical progress, the tidal wave of population growth, the urgent capital demand of the underdeveloped countries, the international defense effort necessitated by

Communist imperialism, the capital requirements of housing (which are artificially enhanced by rent control), and many other things. This list also gives us the reasons which today completely invalidate the fear of inadequate growth potentials, a fear which the economic theory inspired by Keynes has found so difficult to shed during these last twenty years. Our fear is the opposite one. Our problem is how to contain the extraordinarily vigorous forces of growth and how to make sure that the vastly increased capital demand is satisfied by genuine saving and not by the poisonous sources of inflation and taxes—which sources, incidentally, closely communicate beneath the surface.

This brings us to the other aspect of the imbalance between investment and saving. If there is too much investment on the one side, there is too little saving on the other. There arises the question of whether here, too, special forces favor inflation by impairing the ability and inclination to save. This is indeed so. There can be no doubt that in looking for these forces we are on the track of one of the essential causes of the historic uniqueness of today's chronic inflation.

Every act of saving diminishes the pressure of demand on available supplies. It therefore has a dampening, relaxing, and cooling effect on the inflationary boom. To the extent that investment is financed by saving, the critical point of overinvestment is shifted upward, and the boom can continue for a longer period without reaching the danger line. With large saving, there is less need for the central bank to counteract inflation by credit restrictions. Large saving is the most effective means of neutralizing any inflation, whether it is due to overinvestment or to another cause. Nor is this all. Large saving and a large number of savers imply the existence of broad circles of persons and institutions whose attitudes and interests cause them to support sound money. This, in turn, means a broadening of the anti-inflationary front, without whose pressure no active fight against inflation can be expected from the central bank or the government.

Thus saving, which got such poor marks in the theories inspired by Keynes, is again assigned the place of honor which common sense always regarded as saving's due. How to maintain and promote saving, and a host of problems concerning the quantitative effect on saving of different circumstances, laws and taxes—these are the questions which today determine the whole complexion of economic, financial, and social policy. It is also becoming clear that the "negative" saving of consumer credit may well cause the situation to deteriorate considerably.

The question of how to promote saving becomes the more serious by reason of the fact that saving is an activity which an economy and society based on free impulses can least take for granted and rely upon. Yet the very freedom of economy and society depends upon sufficient voluntary saving. The motives which induce people to save are not nearly as dependable as those which cause them to produce, invest, and consume. Keynes and his school quite rightly pointed out that it is possible to go ahead with producing, investing, and consuming without saving at the same time, but they neglected to make sufficiently clear the price which must be paid, namely, inflation and loss of freedom.

In any event, the motives for saving are highly sensitive and vulnerable, and very little would probably be saved if nature and society had not created very strong inducements to save, such as illness, debility in old age, uncertain expectation of life, and the institution of the family. These inducements to save are a powerful support for our economic and social freedom, but they can be systematically destroyed if one only tries. This is precisely what we are busy doing nearly everywhere now, in the age of the welfare state and loosening family ties, forgetting that we are thereby chopping away at the very roots of our free society and economy. To say it briefly, today's super-state, with its super-budget, super-taxation, and super-welfare programs, has developed into a colossal apparatus for dissaving and, at the same time, an apparatus of inflation and growing compulsion. To close the vicious circle, this same inflation, which is due to insufficient saving, gravely impairs further saving because it shakes the saver's confidence in the stability of his savings' value.

This development is well epitomized in the story of an old miner which a pit manager from the Ruhr told me recently. The old man had put aside a tidy sum for his and his wife's old age, but suddenly he decided to blow it all and to spend his little fortune on a luxury television set and other things. Surprised, the manager asked him why he had suddenly changed his mind and was spending all his savings; the old miner replied that the welfare state was now taking care of him anyway and there was therefore no reason why he should deny himself the immediate enjoyment of what he had set aside for his old age.

Two things are now clear. First of all, it is plain that inflationary pressure and saving are like force and counterforce. Processes which weaken the counterforce at the same time strengthen and cause inflation, and these causes are highly effective nowadays. But it is also plain that the obstacles to saving which are among the causes of today's inflation were rarely, if ever, so widely effective before. Indeed, some of them were unknown. We have found what we were looking for; a cause which is historically as new as today's chronic inflation. The causes of the diminution of saving are indeed novel; they have never happened before; they are "modern," as the advocates of super-state, super-taxation, and super-welfare state will no doubt tell us proudly. Another thing is worth remembering: this new thing by which we can explain the permanent inflationary pressure of our age as new in its turn is the result of profound moral and social changes, and of changes which must be regarded as pathological, if only because their ultimate effect is a disease of money, the "democratic-social" inflation of our age.[30]

However, obstacles to saving are only one of the aspects which we have to consider if we want to understand that today's chronic inflation has no historical precedents and must therefore be explained by equally unprecedented causes. We shall encounter the other and at least equally important of these historically unique inflationary forces as we turn to wage inflation, the last of the four sources of inflation which are operative today.

Wage Inflation

The meaning of the term "wage inflation" is no doubt more or less clear. The labor market is the source of continuous new doses of inflation because wages are pushed up so high that the balance between money and goods is disturbed. This happens when the increase in demand because of higher wages is not matched by a corresponding increase in the supply of goods, or in other words, when it is not justified by a corresponding increase in productivity. The result is inflationary excess demand.

It follows that not all of the wage increases to which we have become accustomed nowadays are inflationary. The productivity of labor is certainly rising because of technical progress, expanding investment, and organizational improvements, and we should therefore normally expect a certain increase in the wage level. The continually rising price of labor would be a pure blessing if the productivity increase which higher wages reflect were more or less evenly distributed over all kinds of labor. But in actual fact, it is strongly concentrated on mechanized mass production, while other branches, though affected in varying degrees, lag far behind. This is what causes the problem. The high wages which are possible and natural in industrial mass production and which, thanks to the spreading of automation, promise to go on rising become determining for the overall wage level. The result is the kind of structural prosperity which we see in the United States; it has its advantages, but it also has its serious problems.

The most striking result of this process is that the articles of industrial mass production—the gadgets and technical marvels, from automobiles to television sets—are more inflation-proof than others. They may even become cheaper as other prices are forced all the higher. Certain goods and services whose productivity increases fall markedly short of those current in industrial mass production feel the impact of the rising labor costs generated by the latter. The process seems to be irresistible. Handmade and individual-quality products, personal services, anything made to measure or according to individual tastes, anything produced without stop watch and assembly line and now automation, anything not made to the same last—all of these become more and more expensive and finally exceptional, like a book of unusual quality which can count on only a limited number of readers. It is easy to forget that the prosperity of the automobile and radio is flanked by a decline in all of these other spheres, by impoverishment and shortages which depress the real value of those same high labor incomes which cause them. That our barber will raise his prices next year is almost as certain as tomorrow's sunrise, even without other contributing factors, that is, even if wage inflation were not at work.

Further consequences follow. One of them is that people begin to "do it yourself" in their spare time instead of calling in the prohibitively expensive skilled workman. Like Tom Sawyer, we may persuade ourselves and others that whitewashing a fence is a rare pleasure and a privilege, and, of course, as long as this return to self-sufficiency does not become a drudgery, it is that much progress and gain. Otherwise we have the charming situation in which there are two different wage levels set side by side: the open "American" of high money wages and the secret "Japanese" of our domestic drudgery, with its necessarily low valuation of our own work and time.

Another consequence is the certain prospect that important services like building, nursing, and catering, as well as some staple products whose production cannot easily be mechanized, will go on rising in price until a substitute is found. An example is European coal. Contrary to American coal, European—and especially Ruhr—coal has to be mined in such geological conditions that mining can be mechanized only within limits which cause its productivity to

205

lag considerably behind the productivity level of industry. In this case there is the additional factor that labor is less and less attracted to mining, so that mining wages actually have to exceed industrial wages. As a result, we shall have to face a long-term rise in coal prices until we have sufficient other sources of energy, even without simultaneous wage inflation.[31] Alternately, if cheaper sources of power actually do become available, we may have to face a general depression in coal mining such as is already in evidence in Germany and Belgium.

One last important consequence deserves to be mentioned in this context. It is that agriculture occupies a special position. In spite of mechanization and rationalization, agricultural labor remains, on the whole, manual, at least in Europe, and when, as in the United States, an attempt is made to turn it into a kind of mechanized factory labor, nature sooner or later takes her revenge. Thanks to the determining influence of industrial wages pushed up by mass production, agricultural products would, like those of the craftsman, become steadily more expensive were it not that, unlike the latter and all other services, agricultural products have to face competition from the rest of the world, where production conditions are usually entirely different. Take the European example again. In spite of tractors and milking machines, agricultural productivity falls far short of the industrial level and costs rise continually; yet there is little elbow room for compensating price rises. At the same time, the industrial and urban consumers have a justified interest in low agricultural prices, and agriculture cannot, therefore, widen its elbow room by keeping off foreign competitors. This is the real cause of the difficulties, grievances, and worries of European farmers. The best that can be expected is that the cost increases due to industrial "prosperity" can be offset by rationalization of agriculture to such an extent that only a tolerable residue remains to be offset by protective measures.

All of this, we repeat, would happen also in the absence of inflationary influences. Productivity—or, to be theoretically correct, marginal productivity—is a legitimate and decisive determinant of wages, and international wage differentials are undoubtedly, in the last analysis, due to differences in productivity, which, in turn, is essentially determined by the capital intensity of production. It is therefore normal, natural, and in accordance with the elementary laws of economic theory that a country's average wage level should rise when productivity rises. Nothing is further from the truth than that this must necessarily have inflationary consequences. Does this imply that we shall be safe from the inflationary effects of wage increases as long as the latter do not outrun simultaneous productivity increases? Do we have some kind of safeguard in a field otherwise apparently dominated by power, arbitrariness, or vague claims for justice? We have every reason to be doubtful, if only because, in all of the examples by which we have illustrated the long-term tendency of rising prices for individual services, the influence of a general increase in productivity can, in fact, not be separated from inflationary price rises.

No one would deny the validity of the proposition that wage increases must be matched by corresponding productivity increases in order not to have inflationary effects. This is the least we must insist upon. But if we think a little more carefully about the relationship between wages and productivity and if we try to apply the above formula in practice, we see that it is inadequate and open to dangerous abuse. While the parallelism of wage and productivity increases may safeguard us against the worst excesses of wage policy, it does not offer any guarantee that wage increases will not have inflationary effects. Is it economically rational and does it correspond to the nature of the market economy, to allow all productivity increases to be taken up by higher wages, since these productivity increases are, after all, largely due to technical progress, improved production methods, and higher capital input? Is it not right that productivity increases should also benefit the firm and the consumer, the former by higher profits in proportion to its current and future capital and the latter by lower prices? And if wages are really to rise to the extent of productivity increases, how are the latter to be calculated? And how is it possible to prevent the maximum productivity increase in mass production industries from unjustifiedly pulling along the wages of other branches with smaller productivity increases, down to the crafts, catering, and agriculture? Here, as everywhere else, we are getting into the habit of thinking in statistical terms, but is this at all compatible with the nature of the market economy? Are we not going dangerously astray with our "market economy by statistics"? Where is the boundary between what is normal, natural, and in accordance with the laws of the market economy on the one side and its inflationary violation on the other?

The process of a wage rise due to a productivity rise is normal and healthy as long as it comes to pass by means of the forces of the market and not by a combination of social power and statistics. What happens in the former case is that a wage increase in one of the industries where productivity rises most—say, the automobile industry—spreads to other industries and sectors, not because productivity statistics are used as an argument, but under the influence of supply and demand on the labor market. The primary wage increase attracts workers from elsewhere, and the actual or threatened loss of workers causes wages to rise also in branches where productivity has risen only slightly or not at all. At the same time, wage movements in the two spheres of the primary and secondary wage increase will level out; since the industries where wages rose first because productivity rose most attract labor, wage increases there will be damped as much as they are promoted elsewhere. By its very propagation, the primary wage increase will automatically lose some of its impetus, if only we allow the labor market to behave as we expect all other markets to behave.

Three consequences follow. First of all, the productivity increase (which we assume to have generated the process) will, rationally and equitably, be translated into wage increases only in part, while another part will be taken up by price reductions or by higher

208

profits for the capital which enables productivity to rise. Secondly, the primary wage increase will spread, in gradually diminishing waves, to other sectors of the economy and so cause the productive resources to concentrate slowly but steadily on those branches where productivity is at or near the maximum. On the further condition that the central bank does not continually expand credit and so relieve the labor market of the responsibility for unemployment due to exceeding the equilibrium wage rate, no inflationary effect will follow from the wage increase due to the original productivity rise and from the propagation of that wage increase.

It is natural, inevitable, and of no danger from the monetary point of view that a country's average wage level should reflect the general level of productivity (which is largely determined by capital intensity) and should follow its rises. But it is unnatural, avoidable, and very dangerous for the currency when the connection between wage level and productivity is established, not by the market, but by recourse to productivity statistics, which in any case are invariably doubtful. More and more often in our days the statistically calculated productivity rise of the most favored industries is taken as justifying a claim for a corresponding wage increase. In that case no productivity increase can prevent the wage increase from becoming inflationary, and, unhampered by the laws of the market, the productivity increase will be fully translated into a wage increase instead of leaving room also for a price reduction and profit rise. If we replace the laws of supply and demand on the labor market with productivity statistics and give the latter all the weight of the monopolistic trade-unions' social power, then we tread a dangerous path. It is superfluous to comment on a wage policy which is not even based on statistics but plays only the trump card of social power.

Unfortunately, it is becoming the rule today that wage policy, at best, keeps abusing the argument of productivity increases and more often than not neglects it altogether. In these circumstances it is the most dangerous source of inflation.

The principal reason why wage inflation is so dangerous is that it creates its own conditions of cumulative development. As the inflationary boom proceeds, full employment reaches the stage more properly defined as overfull employment. In a growing number of industries job vacancies exceed the number of suitable applicants, and this state of affairs cannot continue without further inflationary pressure. Indeed, it is of the essence of inflation.[32] Even in the absence of trade-unions, the excess of demand over supply on the labor market would necessarily push up labor costs. This is a dangerous tendency as such and was not unknown in previous boom periods; what is new is that in all countries this tendency is now strongly reinforced by trade-unions, whose power seems to be limited by nothing except their sense of responsibility. Thus a cost inflation due to overfull employment is reinforced by the monopoly power of trade-unions. At the same time, government and central banks find it much more difficult to stem the tide of this inflation by means of restrictive monetary and credit policy.

Another danger of overfull employment is that it may set off a wage-price spiral in which rising wages and prices keep pushing each other up, especially and most effectively in the presence of the fatal system of a sliding scale of wages determined by the cost-of-living index. However, it would be wrong to imagine that government and central banks are powerless in the face of this mechanism. The truth is that the wage-price spiral presupposes continual injections of new money. Otherwise employers would be unable to pay the higher wages without dismissing workers, and consumers would not have the purchasing power to buy the previous amount of goods at higher prices. Unless, therefore, the wage-price spiral is continually supported by the authorities which control the volume of monetary circulation, the wage increases due to overfull employment would inevitably render part of the labor force unemployed. It follows that when an economy gets into the stage of overfull employment (in the precise sense defined above) which is peculiar to chronic inflation, further wage increases become almost

210

inevitable because of the combination of overfull employment and trade-union monopoly, and as a result, the economy finds itself faced with the grave alternatives of inflation or unemployment. Normally, excessively high wages should have the same effect as any other excessively high prices, namely, to render part of the labor force unsalable (as the Americans say, "labor pricing itself out of the market"). In other words, workers would have to be dismissed. If the authorities responsible for monetary circulation wish to prevent this, they must tolerate credit expansion, or loosen credit restriction, to the extent of allowing the inflational wage-price spiral to go on. Monetary and credit policy is reduced to a continuous race between wage increases, which are a potential source of unemployment, and inflationary credit expansion designed to cancel this potential effect of each new round of wage increases.

We have now touched upon the central point of the whole discussion about present-day chronic inflation. We are in a situation in which full employment and rising wages can no longer be combined without inflation. In other words, we cannot have all three: stable money, full employment, and further wage increases. We have to sacrifice one in order to preserve any combination of the other two: stable money and full employment without further wage increases, or stable money and wage increases without full employment, or full employment and wage increases without stable money. Those who now insist upon an "expansionary" or "dynamic" wage policy under any label must accept a steadily progressing crumbling of the value of money and, indeed, the blame of being principally responsible for it. They are the most striking prototypes of all those thousands who keep complaining about inflation and hold others responsible for it but who, at the same time, raise and support claims which make inflation inevitable.

Some trade-unions still take the trouble to base their wage claims on productivity increases. We have already pointed out that this argument is frequently abused. Moreover, even the most favorable price and wage statistics hardly lend themselves to supporting further wage claims now. But the decisive objection is that as long as overfull employment prevails and as long as trade-unions keep the monopoly powers they now possess in most industries, wages are bound to rise above the level justified by productivity increases. With overfull employment, the labor market is a sellers' market, and even a wage rate determined by free competition is bound to be inflationary; much more so, of course, a wage rate to which trade-union power imparts a monopolistic element and which therefore exceeds the theoretical competitive wage rate. Surely, just when trade-unions operate in a sellers' market, they can hardly be expected magnanimously to forego exploiting such a unique opportunity for monopolistic price policy. Still less can they be expected to use their power in order to prevent wages reaching even the level corresponding to free demand and supply in a situation of overfull employment, a level which in itself has certain inflationary effects.

It is obvious that the only thing to be expected of trade-unions in this situation is that they will go on making it worse. Nor can we look to the employers for any real improvement. How are they to put up serious resistance against wage claims in a situation of overfull employment when, moreover, continuous credit expansion—or insufficient credit restriction—enables them to pay a larger wage bill without much detriment to profitability and without dismissing workers? In addition, wage inflation and investment inflation are so closely linked as to be almost inseparable because both draw in the same manner on the source of expansionary, or insufficiently restrictive, credit and monetary policy. If anyone thinks that he stands to gain by investment inflation (or by any other type of inflation, including imported inflation), he will have to take wage inflation into the bargain and vice versa. There is little sense in passing the ball of responsibility backwards and forwards. The beneficiaries of wage inflation and investment inflation (or any other form of current inflation) eventually have common interests, which are concealed behind the bargaining on the labor market. Nevertheless, the real seat of resistance, the hard core, the disturbing and intractable novelty, is wage inflation and the chain of its effects.

Nor, on reflection, is there any hope of escaping the wage-price spiral and the dilemma it creates by means of increased investment designed to offset the higher wages with higher productivity. This would be adding fuel to the flames; far from avoiding wage inflation by means of accentuating investment inflation, both would be aggravated.

We conclude that no one can relieve the authorities responsible for monetary and credit policy of the dilemma. It is a heavy burden; just how heavy becomes clear when we consider that the country concerned is by now caught in a whirlpool from which it can escape only if a sufficiently restrictive monetary and credit policy reduces overfull employment to normal full employment so that the labor market ceases to be a sellers' market and the buyers gain a slight predominance. But once the country is caught in the whirlpool of inflationary overfull employment, this can hardly be done without putting up with some temporary unemployment as a price for winning the battle of inflation. It is in this light that we must understand the recent remark, made by a British expert, that it is "almost certain that stable prices cannot be maintained [in Great Britain] at a level of demand that keeps unemployment much lower than 2 per cent."[33]

Wage inflation, then, can be stopped effectively only if the central bank refuses to play the game and restricts credit until the chain reaction of overfull employment and trade-union power, wage increases, price increases, and more wage increases is broken. The more restrained trade-unions are in exploiting their strong bargaining position, the smaller is the required dose of credit restriction. On the other hand, the longer the central bank waits until it turns off the tap of credit and the further it lets matters drift, the more difficult the operation becomes because the amount of temporary unemployment involved would then be much more painful.

The decision is always difficult. At present, it is made even more

213

difficult by another circumstance, which is also a novelty in our age and which we must also take into account if we want to understand the uniqueness of our chronic inflation. Monopolistic trade-union power is new, and so is the disinclination to save; the third novelty is the overriding and dogmatic conviction that "full employment" is to be valued above all else and allows no compromise—not even a normal measure of dispersed and temporary unemployment having nothing in common with the terrible mass unemployment of the Great Depression. Governments and central banks feel constrained by public opinion to accept this dogma, even if they are not, as in the United States, obliged by law to do so. At the very least, this dogma holds such sway as to have prevented any government or central bank, so far, from seriously attempting to interrupt the chain reaction. The pressure on the authorities is further augmented by the fact that it has become part and parcel of the full-employment dogma to deny the above-mentioned relationship between wage level, inflation, and unemployment and to regard as the outcome of perverse economic theory the necessary conclusion that there must be a certain, however small, number of unemployed.[34] We end up with the fatal attitude of viewing every slight increase of unemployment as a failure on the part of the central bank and of blaming it for neglecting its guardianship of full employment. The truth is, of course, that the central bank has to defend the purchasing power of money against the consequences of a reckless wage policy. But as long as this attitude prevails, there is a constant danger that any determined effort to suppress wage inflation will be called off again at the first sign of the slightest fall in employment, whereupon the spiral of wages and prices is once more given free play.

In discharging its responsibilities, the central bank is not unlike a motorist who knows that by pushing hard on the brake pedal he can prevent an accident but who also knows that pedestrians are inconsiderate and undisciplined and that the road is so slippery that he might lose control of his car if he braked too sharply. Both are mitigating circumstances. Nevertheless, we expect him to remember his brakes and not to rely on honking and shouting. The sooner he recognizes the danger and operates the brakes, the better.

This is an accurate enough picture of the unenviable position of the central bank. It certainly has the ultimate responsibilty. But overfull employment, ruthless exploitation of trade-union power and the simultaneous inflationary impulses from elsewhere threaten to require more from the exercise of this responsibility than public opinion, government, and the full-employment dogma are prepared to accept. In these circumstances the independence of the central bank is invaluable, and so is a management which intelligently and vigorously takes advantage of its independence.

It should now be clear what is meant by the statement that the present-day inflationary process is such that the power of the central bank hardly suffices to control inflation. We see that this must be understood correctly. It does not mean that a central bank equipped with all the tools of modern credit policy, so that it does not need to rely solely on the blunted weapon of discount policy, is unable to eliminate excess demand, from whatever source, by appropriate restrictions. There is no doubt whatever that the bank of issue can do this, now as before. The real and disturbing question is whether the central bank can apply the brakes sufficiently strongly, because the ensuing fall in employment, which is almost inevitable in view of the trade-unions' wage policy, will no doubt call forth such political and social resistances that the bank's action is paralyzed or even nipped in the bud at the stage of decision.

There is yet more to it than that. Wage inflation is the most important and intractable form of inflation but, as we have seen, not the only one. All the other sources of inflation are active at the same time, reinforce each other, and combine with wage inflation. This is true not only of investment inflation, which is so intimately connected with wage inflation, but also of fiscal inflation and, in the particular case of Germany, imported inflation. Given the ubiquity of inflationary pressure, we may be justified in wondering

215

whether it is not asking too much of the central bank to neutralize all of these inflationary impulses with the shock absorber of credit restriction. This is, of course, what theory requires, and in fact the central bank should do everything in its power. But whether this is enough remains an open question.

Three other measures are imperative. Every effort should be made to induce the trade-unions to refrain from exploiting their monopoly position to the full by appealing to their reason and sense of responsibility, backed by the increasingly obvious damage which the race between wages and prices inflicts on the workers' own interests. At the same time, fiscal inflation must be stopped. It must be stopped by cutting down government expenditure and not by the dubious means of higher taxation, which would be contrary to the long-term goal of reducing the elephantiasis of the budget and eliminating fiscal socialism.[35] Thirdly, any inflationary sources due to balance-of-payments surpluses must be stopped up. For the rest, we repeat emphatically, the bank of issue must, all difficulties notwithstanding, energetically restrict credit until investment has been reduced to the—we hope rising—level of saving and wage claims to a level at which they do not cause wage inflation. If the trade-unions should persist in making excessive claims, they will eventually find that employers are no longer able to raise wages, and they will therefore have to face the choice of yielding or of putting up with unemployment.

Conclusions and Prospects

What conclusions are we to draw from all this? First of all, surely, that in inflation the defense must fight on as broad a front as that of inflation's causes; furthermore, the heavy artillery of the bank of issue must occupy the center and be used with the maximum of firepower. But the broadening of the front is not enough; the battle against so tough an enemy, pressing in from all sides, is bound to be long and fluctuating and must be fought in depth as well.

The main thing is the fighting morale of the defense. We are obliged to admit that this is anything but brilliant. To say this is to say once more that the chronic inflation of our age is a moral and social problem. For this very reason it is also a problem which is not susceptible to any comfortable, quick, and simple solution or to one of mere monetary and credit techniques.

The right way of looking at this problem is to regard inflation as an economy's reaction to a continuous and multiple strain on its resources. It is a reaction to extravagant and impatient claims; to a tendency toward excess in all fields and among all classes; to inconsistent and confused economic, financial, and social policies which disregard all time-tested principles; to the presumption of taking on too much at one time; to the recklessness of always drawing more bills of exchange on the economy than it can honor; to the obstinacy of always wanting to combine what cannot be combined. People want to invest more than saving allows; they claim wages higher than those corresponding to the rise in productivity; they want to consume more than current income can pay; they want to earn more with exports than the latter can yield the economy by way of imports; and on top of all this, the government, which should know better, keeps extending its own claims on the economy's strained resources. Demands proliferate while the necessary cover of goods is missing. If any man should continually sin against all the rules of reasonable living, some organ of his body will slowly but surely suffer from the accumulation of his mistakes; the economy, too, has a very sensitive organ of this kind. This organ is money; it softens and yields, and its softening is what we call inflation, a dilatation of money, as it were, a managerial disease of the economy.

The trouble is that we lack counterforces of a spiritual, moral, and social nature. In the realm of ideas we no longer have definite

217

convictions and guiding principles; in the realm of interests the anti-inflationary front is neither strong nor broad enough to meet the inflationary one, and, as we have seen, the welfare state makes gaping breaches in the front of those who have a vital interest in suppressing inflation. Wages and pensions geared to the cost-of-living index do the rest. The result is that respect for money and its inviolability has lost its force.

To show how far we have already gone in this direction, let me tell two stories from the financial history of France.[36]

At the close of 1870, in one of his country's gravest moments, Gambetta was organizing the Republic's resistance at Tours. He requested the local representative of the Bank of France to help him in his desperate need of money by printing bank notes. At the time, the request was an enormity, and in fact, Gambetta, Jacobin firebrand though he was, as well as an all-powerful dictator, had to give way to the flat refusal of a banker who would not accept even an extreme national emergency as an excuse for the crime of inflation. Gambetta eventually succeeded in raising a loan of two hundred million francs, which Morgans lent him at 7 per cent.

The second story is perhaps even more impressive. A few months later, in March, 1871, the leaders of the Commune in Paris were tempted to lay their hands on the gold reserves and printing plates of the Bank of France and thus finance their revolution. But in the midst of a pitiless civil war, these hard-boiled revolutionaries withstood the temptation. The Bank of France was as inviolable to them as was the *franc Germinal*—both the creations of the tyrant Napoleon, who had endowed them with inviolability, although he was otherwise not in the habit of limiting the power of the state in any way.

Both of these incidents reveal that money used to be sacrosanct in the very same country whose governments and parliaments have since shown a most spectacular disrespect for money. Any comparison with our times must fill us with consternation and shame.

We cannot help being touched, impressed, and troubled in reading these stories of Gambetta and the Commune, and if we are honest with ourselves, we must admit that this deference of the *raison d'état* to the inviolability of money was a valuable spiritual asset. We can but lament its loss and admit that it throws us back into the age of the princely forgers of the late Middle Ages and early modern times.[37]

Much has contributed to the undermining of the respect for money and its value. Immeasurable damage has been done, above all, by that revolution in economic theory whose disastrous effects we have already stressed. Its destructive work can hardly be undone now, but we are at least entitled to expect repentance and frank confession. Let us hope that the time will soon come when Keynes, to use one of Jacob Burckhardt's expressions, will be recognized as one of the great intellectual ruiners of history—like Rousseau and Marx.

We are faced with a phenomenon of intellectual and moral decadence, and we meet it in various degrees and forms. It ranges from limp resignation to the crumbling value of money to a mixture of regret and scarcely concealed satisfaction and, ultimately, to open cynicism. In fact, it is just foul play, but this is now often forgotten.

It is nothing short of monetary cynicism, and thoroughly contemptible, to regard the slowly advancing chronic inflation of our times as both possible and desirable, in the sense that it is the price which people ought to be prepared to pay for an unlimited duration of the boom, with overfull employment, continuous wage increases, and "economic growth." This argument of "simmering permanent inflation" is not only morally objectionable and decadent, it is also logically untenable. Inflation is precisely one of those things which cannot be conceived of as permanent because once it is recognized, it loses the leisurely pace which must be a feature of constant and "controlled" inflation. Once people suspect more generally—as they must do sooner or later—that we live in an "age of inflation," they will increasingly behave in such a manner that an explosive chain reaction is set off, to use the terminology of nuclear physics. Cost-of-living clauses will be more widely introduced, and they will effectively accelerate inflation. To maintain overfull, or even only full, employment will eventually require increasingly stronger injections of inflation poison in order to offset the unfavorable effects of rising wages on employment. At the same time, progressing inflation will diminish saving and raise the rate of interest, and ever more inflation will be needed in order to maintain investment at the level necessary to prevent the collapse of the boom. In the end, only the complete destruction of the currency will be able to put off the final crisis for one more, short, period of grace.[38]

It is bad enough that this monetary cynicism is foolish. It is much worse that it is morally objectionable. It is part of the cynicism to take this designation with a smile. If we want to go to the roots of the chronic inflation of our times, then we must recognize that the mental attitude which generates inflation, tolerates it, resists it but feebly, or defends it cynically is the monetary aspect of the general decline of the rule of law and of respect for the law.

Democracy, as we have seen, degenerates into arbitrariness, state omnipotence, and disintegration whenever the decisions of government, as determined by universal suffrage, are not contained by the ultimate limits of natural law, firm norms, and tradition. It is not enough that these should be laid down in constitutions; they must be so firmly lodged in the hearts and minds of men that they can withstand all onslaughts. One of the most important of these norms is the inviolability of money. Today its very foundations are shaken, and this is one of the gravest danger signals for our society and state.

I would remind those of my readers who know it of Jeremias Gotthelf's story *The Black Spider*. For the benefit of those who do not know it, I should explain that it is a tale in which reality and legend interweave in the sinister power of a devilish spider which

220

unleashes the disaster of the plague in a peaceful world, until eventually people manage to lock it into the beam of a farmhouse in Bern by means of a plug.

This black spider is the symbol of an infernal force irrupting into a peaceful society. The symbol may very well be applied to inflation. Inflation, too, has, throughout history, set out to plague us—until it was finally locked in by the plug of the gold standard. But in the despair of a devastating deflation, the nations pulled the golden plug out a quarter of a century ago. Since then, the black spider has been abroad again.

We know that such views will be met with a disdainful smile. It is out of place. We shall be told of the reasons which allegedly make a gold standard impossible today—but what do they all boil down to, ultimately, except that we simply do not have the determination and the strength to tame inflation once more? Are the conditions for a successful taming of inflation different from those of replacing the golden plug?

221

CHAPTER V

Centrism and Decentrism

The Dividing Lines in Social Philosophy and Economic Policy

I return once more to the subject of inflation, which we have just examined from all sides, in full awareness of its overwhelming importance. I would like to relate an experience I had a few years ago. It so happened that on one and the same day I came across two statements, both of which concerned money but which arrived at such completely divergent and indeed irreconcilable conclusions that they were explicable only as being derived from two opposing social philosophies. One was made by a distinguished American economic expert. The very title of the article which contained it was provoking: "Inflation or Liberty?" It was an early warning of the danger of progressive inflation, the kind of warning which must now be recognized by all as justified. The author came to the conclusion that a nation can preserve its freedom only with the help of sound money but that in a modern mass democracy the monetary system could not remain sound if it was at the mercy of government, parliament, political parties, and powerful pressure groups in the absence of sufficient countervailing forces. The other statement reached me through a German news agency. A university professor of moderate socialist tendencies violently criticized the "fatal deflationary policy" of the German central bank and demanded that "the democratic means of guiding the economy, namely, money and credit, should be placed in the hands of democracy."

It is unlikely that this socialist would repeat today what he then propounded so passionately. The American, on the other hand, would have every reason not simply to warn but to implore. It follows that the socialist was wrong on a matter of economic policy, while the American was right. The inflationary pressure which the latter feared has now become so obvious that even the socialists must subordinate all other considerations to the need for an effective barrier to inflation.

But this is not what interests us in this context. The point I wish to make is the sharp and irreconcilable clash of two opinions on one of the most important questions of the economic and social order. It is difficult to think of a compromise between the two principles here involved. Either it is right and desirable that monetary and credit policy should be operated like a switchboard by a government directly dependent upon a parliamentary majority or, worse still, upon some non-parliamentary group posing as the representative of public opinion. Or, conversely, it is right and desirable to counteract such dependence. Either it is wise to put all the eggs into one basket or it is not. It is perfectly obvious that the German professor's opinion, which rested on a characteristic though, as we know today, unfounded fear of deflationary policies on the part of the central bank, was as firmly grounded on profound social convictions as was the American economist's contrary opinion.

We are in the presence of a case which reveals a fundamental cleavage of thought within social philosophy and economic policy. It teaches us how important it is to mark the dividing lines clearly in the everyday strife and conflicts of political opinion. The better we succeed in doing so, the more we may hope to understand the meaning of political differences and to reduce the conflict to an honest and generally recognized opposition of basic convictions. Not the least of the merits of such an undertaking is that it leads us to examine our own conscience and make our own choice. What are we, really? Liberals? Conservatives? Socialists? And if we are one or the other, why? And whither does it lead us?

Our example suggests that we may begin with a contrast which, while not the most important, is closely connected with other and more profound contrasts. We might say that a man whom we would call an inflationist is here in conflict with another whom we might call a deflationist. There is some justification in putting things in this manner. Each of us obviously tends, by temperament, either to think that inflation is a lesser evil than deflation or vice versa; or, in other words, either to fear deflation more than inflation, or vice versa; or, in yet other words, to be quicker in recognizing the dangers either of inflation or deflation. The reader will, at this stage of the book, not be in the dark about the author's views. I would go so far as to deny the justice of calling anyone a deflationist in the same sense in which his opposite number may be called an inflationist, for the simple reason that, as we know, there exists an asymmetry between inflation and deflation. Inflation is a poison whose initial effects are mostly pleasant and which reveals its destructive powers only later, while deflation is a process of general disadvantage from the outset; for this reason it is possible to want inflation, but deflation can at best be accepted as a possibly lesser evil. Hence we may speak of inflationism as an attitude not only of defending but possibly desiring inflation, and it is, in fact, a powerful and ancient current in economic history. Nothing of the kind can be said of deflationism.

To examine inflationism and deflationism in detail and to analyze their motives is a task which is as rewarding as it has hitherto been neglected. Let us look at some of the most important factors. First of all, inflationism has an exaggerated predilection for continuous growth, for rising figures (including the population figure), for quantitative progress—in short, it has a tendency to make sacrifices to expansion within limits which are too widely drawn. We are again reminded of Faust, in his old age this time: this kind of expansionism wishes, as he did, to "see such a throng" and "furnish soil to many millions." It rejoices in steeply rising curves and is prepared to pay for them by letting the curve of the value of money go down, at least for longer than is safe. Such expansionism implies very many other things, too: if necessary, it is willing to sacrifice the more remote future to the present, whether its adherents say, with the eighteenth century, "après nous le déluge" or, in Keynes's more modern terms, "in the long run we are all dead"; it has no feeling for those invaluable reserves of society which ought not to be used up as fuel to keep the boiler of expansion going, including one of the most precious reserves of all, namely, respect for money and the inviolability of its value; it is against anything bourgeois, against the creditor, against the *rentier*, to whom, like Keynes, it wishes at best a painless death. Expansionism is futuristic, optimistic, and much else besides; the deflationist, or, as we would prefer to say, the anti-inflationist, is the opposite in all respects.

At this point of our argument we are anxious to probe deeper. We want to resolve the contrast between inflationism and anti-inflationism into a wider and more general one so that we may gain a vantage point from which to view the ultimate conflict between two social philosophies and two currents of economic policy. I have in mind the conflict between "left" and "right" in thought, between a tendency towards what I called "progressivism" and discussed in detail in my earlier book *Mass und Mitte* and one towards "conservatism"—although I hardly like to use this expression nowadays because in most countries of the Western world it carries undesirable associations.

In order to make the transition from our example to this more general and important dividing line, we recall once more that our unhappy socialist, expansionist or inflationist as he was, objected to the independence of the central bank and demanded that it be subjected to the will of "democracy." Our American, on the other hand, who was convinced that inflation is a danger which always

225

lurks in the background and now threatens us immediately, felt quite certain that a barrier must be erected against government domination of money. This was precisely the essence of the gold standard; after its demise, the only barrier remaining is the independence of the central bank, and it must be defended all the more obstinately. One of our spokesmen wants to concentrate responsibility for money in the hands of government and to subject it to politics; the other wants a division of power, an articulated system of checks and balances, decentralization, and hence the withdrawal of money from politics.

The first talks of the need to put money, the "democratic means of guiding the economy," into the hands of a government acting at its own discretion and according to a comprehensive plan, so that the government may conduct an economic policy which is called "progressive," guarantee "full employment" and thereby the power of trade-unions, and guide the course of the economy according to the wishes of the "people." In advocating all this, our German professor expresses a certain social philosophy diametrically opposed to that of his American counterpart. In this particular question, as in others, he takes a line common to the Jacobins of the French Revolution and all of their many spiritual heirs. The ideal of democracy is seen, not in a well-articulated state with balanced and therefore mutually limiting powers, but in centralized power of a kind which is unlimited in principle and can in practice be wielded all the more freely as it is supported by the fiction of the sovereign will of the people. This Jacobin excludes or repudiates the idea that, in so far as, like Montesquieu *(Esprit des Lois,* Book XI, Chapter 2), we attach more value to the freedom of the people than to its imaginary "power," democracy can derive nothing but benefit when power, of whatever origin, is split up and its mistakes and abuses thereby limited. A man who looks with suspicion upon a central bank which has not yet become a pliable tool of centralized state power reveals himself as one of those "eternal Jacobins" to whom any manifestation of independence and autonomy is a thorn in the flesh, whether it be the free market, free local government, private schools, independent broadcasting, or even the family itself.[1] Any institutions which still preserve some independence, whether central banks or pension funds or anything else, are, to repeat the simile, so many *Bastilles* which have to be razed to the ground.

Clearly, we are faced by two types of social thought to which most specific conflicts may be reduced without difficulty. We seem to be standing on a ridge from which we have a wide view into the valleys on both sides. Here is the parting of minds. Some are attracted towards collectivity, the others to the members which compose it. The former look at the structure of society from the top downwards, the latter from the bottom upwards. The first seek security, happiness, and fulfillment in the subordination of the individual and the small group to a deliberately and strictly organized community, which, from this point of view, is all the more attractive the larger it is; the others seek these benefits in the independence and autonomy of the individual and the small group. The difference in social outlook closely resembles another difference between two modes of thought: one which has a strange predilection for everything contrived, man-made, manufactured, organized, and intricately constructed, for the drawing board, blueprint, and ruler; and another which prefers what is natural, organic, time-tested, spontaneous, and self-regulating, and which endures through long eras. Still another difference in outlook is connected with this. On the one side are those who believe that society and economy can be reconstructed from above and without considering the fine web of the past. They believe in radical new beginnings; they are reformers inspired by an optimism that is apparently proof against any failure. On the other side are those who possess a sense of history and are convinced that the social fabric is highly sensitive to any interference. They deeply distrust every kind of optimistic reforming spirit and do not believe in crusades to conquer some new Jerusalem; they hold, with Burke, that the true statesman must combine capacity for reform with the will to prudent preservation.

Before continuing with this attempt to characterize the two types of social thought, we must confess that while we are most anxious to find a name for them, the task is a somewhat embarrassing one. We tried to give a provisional indication with the expressions "progressivism" and "conservatism," but we had to desist at once because the label "conservatism" is too discredited, at any rate on the European continent. Even with all kinds of qualifications, tiresome misunderstandings would still be inevitable. We do not get much further by contrasting "individualism" and "collectivism" because this would imply some sort of exaggeration in both cases. Nor are "liberalism" and "socialism" the right words. They have both become indispensable in the vocabulary of politics, but for this very reason they have become blurred by use and have collected so many shades of meaning and associations that they are useless for our purpose, especially since they signify something different in nearly every country.[2] What we need is a terminology which is not only new, fresh, and unburdened by associations but which also characterizes at least some essentials of the great contrast. What we have said so far in this chapter suggests the solution of calling the Montagues and Capulets of our play by the names "centrists" and "decentrists."

It should be clear by now that we are in the presence of two contrary principles which determine and mark all aspects of social life—politics, administration, economy, culture, housing, technology, and organization. If we take both concepts in a broad sense and explore their implications to the end, they will be revealed as two principles which express what is perhaps the most general contrast in social philosophy. Whether our ideal is centralization or decentralization, whether we regard as the primary element in society the individual and small groups or the large community, that is, the state, the nation, and the collective units up to the utopian world state—these are the questions which ultimately constitute the watershed between all the currents of thought and points of view which we have so far confronted with each other.[3]

This is where federalism and local government clash with political centralization. It is here that the friends of the peasantry, the crafts, and middle classes, and the small firm and of widely distributed private property and the lovers of nature and of the human scale in all things part company with the advocates of large-scale industry, technical and organizational rationality, huge associations, and giant cities. This is the moat across which the eternal dialogue goes on: on one side are those who think that the economy is best planned by the market, competition, and free prices and who regard the decentralization of economic decisions among millions of separate producers and consumers as the indispensable condition of freedom, justice, and well-being; on the other side are those who prefer planning from above, with the state's compulsory powers. And so it goes on.

The centrist is none other than the social rationalist, whom we met before. Seen from his central point, the individual is small and eventually dwindles to a statistical figure, a building brick, a mathematical magnitude encased in equations, something that can be "refashioned," in short, something that may well be lost sight of. We know with what optimism our social rationalist views the success of his constructions and refashioning. By contrast, the decentrist, who thinks in terms of human beings and also knows and respects history, is skeptical or pessimistic and in any case bases his arguments realistically and unsentimentally upon human nature. The centrist is doctrinaire, the decentrist undoctrinaire and unideological. The latter prefers to hold on to established principles; he is swayed more by a hierarchy of norms and values, by reason and sober reflection, than by passions and feelings; he is firmly rooted in ultimate and absolute convictions for which he requires no proof because he would regard it as absurd not to believe in them.

We see also that the centrist is what we have called a moralist, a

229

moralist of the cheap rhetorical kind, who misuses big words, such as freedom, justice, rights of man, or others, to the point of empty phraseology, who poses as a paragon of virtues and stoops to use his moralism as a political weapon and to represent his more reserved adversary as morally inferior. Since, again, he looks at things from on high, well above the reality of individual people, his moralism is of an abstract, intellectual kind. It enables him to feel morally superior to others for the simple reason that he stakes his moral claims so high and makes demands on human nature without considering either the concrete conditions or the possible consequences of the fulfillment of those demands. He does not seem capable of imagining that others may not be lesser men because they make things less easy for themselves and do take account of the complications and difficulties of a practical and concrete code of ethics within which it is not unusual to will the good and work the bad.

The "left" moralist all too often reaches the point where his big words of love and freedom and justice serve as a cover for the exact opposite. The moralist, with his lofty admonitions, becomes an intolerant hater and envier, the theoretical pacifist an imperialist when it comes to the practical test, and the advocate of abstract social justice an ambitious place-hunter. These moralists are a world apart from the decentrists' attitude, of which the hero's father in Adalbert Stifter's *Nachsommer* says that man does not primarily exist for the sake of human society but for his own sake, "and if each one of us exists in the best possible manner for his own sake, he does so for society as well." I used to know an old servant who had discovered this wisdom for herself; she always wondered why so many people kept racking their brains about how to do good to others, while, so she thought, it would surely be better if everyone simply and decently did his duty in his own station. The centrist's moral ideal frequently enough amounts to a desire to make the world into a place where, to quote Goethe again, everyone is nursing his neighbor—which presupposes a centralized compulsory organization.

The further we proceed with our analysis of the two modes of thought, the more we are led to assign each attitude to one or the other camp. The contrast between centrism and decentrism is, in fact, unusually comprehensive. In the economic sphere, the contrast is most clearly epitomized by monopoly and competition, and the collectivist economy corresponds to the centrist's ideal, just as the market economy corresponds to the decentrist's. Every economic intervention is a concession to centrism—made lightheartedly and in pursuit of his own ideal by the centrist and unwillingly by the decentrist. The latter demands strict justification for all concessions, and the burden of proof is on their advocates because it is his principle that there is always a presumption in favor of shifting the center of gravity of society and economy downwards, so that every act of centralization and every upward shift of the center of gravity requires convincing proof before the decentrist will condone such deviation from his ideal.

The position of equality and inequality cannot be in doubt. Equality and uniformity obviously belong to centrism; inequality, diversity, multiformity, and social articulation to decentrism. This requires no further explanation, but there is a special problem here upon which we touched earlier (Chapter IV, Note 6), namely, the particular form of "equality of opportunity." This problem reminds us that life is not an equation which is soluble without a remainder; unless we are very careful, decentrism might involve itself in self-annihilating contradictions on this point. The ideal of decentrism, in common accord with one of the unchallenged aims of liberalism, certainly demands that individuals should try their strength against each other in free competition, and this implies that they start the race from the same starting line and on the same conditions. Is it, then, to be a continuous race of all for everything? Do we always have to be on the lookout for better opportunities, wherever they may appear? Do we always have to regret the opportunities we missed and always chase after those we think are better? This cannot be the true meaning of the ideal. If it were so, it would obviously be a dangerous ideal and one most uncongenial to the

231

decentrist, and to pursue it would cause general unhappiness. Our star witness, Tocqueville, observed long ago that the Americans, in whose country equality of opportunity always held pride of place, are so dedicated to the restless hunt after better opportunities that they end up as nervous and ever dissatisfied nomads.[4]

An uncommonly impressive and at the same time repulsive symbol of such a race of all for everything is to be found in the spectacle that memorable day, more than half a century ago, when a part of the territory of the present state of Oklahoma (the land had been taken from the Indians) was thrown open to settlers. They were waiting at the border, and at the shot of a pistol they all rushed forward from this completely equal starting line to compete for the best plots of land. Surely it must be obvious to everyone that nothing could be more unwise or dangerous than to turn society into such a continual race. Even if the production of goods could so be maximized, it would not be worth the price. Men would be incessantly on the move; culture, happiness, and nerves would be destroyed by an unending to and fro and up and down from place to place, from profession to profession, from one social class to another, from "shirt sleeves" to a fortune of millions and back to "shirt sleeves." No, the deeper—we might say here the conservative—meaning of decentrism is that it behooves us to bethink ourselves of the indispensable conditions for a sound and happy society. These are a certain stratification of society, respect for natural developments, a modicum of variety and of horizontal and vertical social articulation, family traditions, personal inclinations, and inherited wealth. From this point of view, it is, for example, by no means foolish if a country's townships or districts try to preserve their character to some extent by not immediately granting every newcomer the same rights as are enjoyed by the original inhabitants.

It is not good if all the sons of peasants and bakers should become, or wish to become, physicians, clergymen, or clerks. It is true, now as always, that it is highly desirable that men should have the happy feeling of being in the place where they belong—indeed, it is truer than ever in our age, when this feeling has become so rare because of the ideal of the race of all against all. Frédéric Le Play, the nineteenth-century engineer and sociologist, was not so stupid when he discovered an important mainstay of society in the *familles-souches*, the families in which profession and economic and social position are handed down from father to son.[5] Finally, it deserves to be stressed that if equality of opportunity is to be achieved by socializing education, envy and resentment will only be acerbated. If everybody has the same chances of advancement, those left behind will lose the face-saving and acceptable excuse of social injustice and lowly birth. The weakness of mind or character of the overwhelming majority of average or below-average people will be harshly revealed as the reason for failure, and it would be a poor observer of the human soul who thought that this revelation would not prove poisonous. No more murderous attack on the sum total of human happiness can be imagined than this kind of equality of opportunity, for, given the aristocratic distribution of the higher gifts of mind and character among a few only, such equality will benefit a small minority and make the majority all the unhappier.

In order not to stray from the right path, we must always remember that the ideal of decentrism requires us to stand for variety and independence in every sphere. However, it would be equally wrong if we were to confuse decentrism with particularism or parochialism and with parish-pump politics—that is to say, with a narrow-mindedness which can't see the forest for the trees. This is not what is meant. The decentrist must in all circumstances be a convinced universalist; he must keep his eye on a larger community which is all the more genuine for being structured and articulated. His center is God, and this is why he refuses to accept human centers instead, that is, precisely that which consistent centrism, in the form of collectivism, intends to present him with. This is how he understands the inscription placed on Ignatius de

233

Loyola's grave: "Not to be excluded from the greatest, yet to remain included in the smallest, that is divine." This, no doubt, is also what Goethe had in mind when he said:

> I am a citizen of the world,
> I am a citizen of Weimar.[6]

The right spirit is one which enables us to combine an overall view with a sense of the particular. On the one hand we should cultivate a universal approach to all intellectual, political, and economic matters and reject narrow views and actions and, above all, intellectual, political, and economic regionalism and nationalism; on the other hand, we should prize variety and independence at all levels and in all spheres, on the basis of the common patrimony of mankind, which is beyond all levels and spheres.

Apart from many other insights which the centrist lacks, the decentrist also knows that it is always easier to centralize than to decentralize and to widen the powers of the state than to curtail them. There is yet another thing which the decentrist knows better, and this is that the centrist's path is bound to lead to regions where the air of freedom and humanity becomes thinner and thinner, until we end up on the icy peaks of totalitarianism, from which nations can hardly hope to escape without a fall. The trouble is that once one takes this road, it becomes increasingly difficult to turn back. Centrism is in danger of encountering no check any more, least of all in itself. The obsession of uninhibited centrism can, like so many other things, be illustrated by a story from the world's store of legends. It characterizes with exaggerated symbolism both the direction of the march and the secret wishes of its leaders. I have in mind the story of Caligula, who is reported to have expressed the wish that the people of Rome might have but a single head so that it could be decapitated with one stroke. Caligula's wish has always remained the symbol of a kind of centrism which is tyrannical because it knows no limits and also a symbol of the inevitable end to which centralization must lead.

The temptation of centrism has been great at all times, as regards both theory and political action. It is the temptation of mechanical perfection and of uniformity at the expense of freedom. Perhaps Montesquieu was right when he said (*Esprit des Lois*, XXIX, 18) that it is the small minds, above all, which succumb to this temptation. Once the mania of uniformity and centralization spreads and once the centrists begin to lay down the law of the land, then we are in the presence of one of the most serious danger signals warning us of the impending loss of freedom, humanity, and the health of society. A century ago, John Stuart Mill wrote: "If the roads, the railways, the banks, the insurance offices, the great joint-stock companies, the universities, and the public charities, were all of them branches of the government; if, in addition, the municipal corporations and local boards, with all that now devolves on them, became departments of the central administration; if the employés of all these different enterprises were appointed and paid by the government, and looked to the government for every rise in life; not all the freedom of the press and popular constitution of the legislature would make this or any other country free otherwise than in name. And the evil would be greater, the more efficiently and scientifically the administrative machinery was constructed."[7]

The Web of Human Relations

The dangers and temptations of centrism are the more considerable because of their great variety. We must always be on guard against unwittingly making concessions to centrism or promoting it against our intentions. The world is full of centrists who neither want to be centrists nor realize what they are; these are the liberals or conservatives who reject federalism, the anti-collectivists who flirt with monopoly or government intervention in the economy, humanistic Europeans who support an economocratic organization of our continent, and many others.

In approaching the end of this book, it should hardly be necessary to point once more to the most striking and well-known symptoms of growing concentration all around us, with which, as a rule, we put up readily enough or to which we even give our blessing. But it may be useful to sharpen our awareness of the dangers of concentration and of the centrist attitude which fosters it. We shall try to do so by means of a few examples from less familiar fields.

Most important in this context is the fact that the web of human relations is growing closer as a result of the steady increase in the number of dependent wage earners. The individual is getting caught in a situation of subordination and dependence in relation to centers of decision. This process is part of the great upheaval which the American economist K. E. Boulding calls the organizational revolution, and it disturbs the balance of human relations. The relation between independent market parties, the buyers and sellers, is horizontal and loose, if not impersonal. As firms grow in size and the number of independent market parties diminishes, the market's more or less impersonal and loose co-ordination is replaced by the vertical, close, and personal relation of subordination and authority. Dependence upon the client or the supplier through a market wide enough to do away with rigid personal relationships is replaced by dependence upon the boss.

People used to occupy positions side by side with each other, but now they are above and below each other, and the relation is charged with the constant tension of close personal contact within a limited, fixed group. With the diminution of individual independence, this is becoming the fate of the masses, and we all know the strain it puts on human relations. Intrigues, place-hunting, informing, ill will, bootlicking, envy, jealousy, and all the other poisons of close contact spread like the plague in all large organizations and companies, as experience has shown again and again. Neurotics are in a position to make life hell for hundreds and thousands of people, and, as Boulding points out, there is a more than even chance that it will be precisely neurotics who get to the top and into a dominating position, because of their assertiveness

236

and officiousness. A bad-tempered tax collector can let himself go with both his subordinates and with the taxpayers at his mercy; the psychologically unbalanced foreman can become the factory tyrant and intimidate all the other workers. But however irritated or worried the greengrocer may be, he has to pull himself together, without, on that account, having to feel that he is the slave of his customers.

The worst is that this organizational revolution also catches up with those who do not as yet belong to some large concern and who still have their professional independence. What, for example, has happened to the medical profession? The life of physicians—especially in highly centralized welfare states—has become utterly exhausting because they are now involved in a situation of double dependence: the old, horizontal market dependence in relation to their patients, and the new, vertical, organizational dependence in relation to the health-insurance funds. It is nothing short of a tragedy that this vice should have gripped precisely the profession upon whose calm and composure our life and health depend.

At first sight these considerations constitute a new and impressive proof of the superiority of the market economy over any kind and degree of collectivist economy. It is impossible to overstate the value of the impersonal integration of people through the market in comparison with their conglomeration in a collectivist economy, however much the former may be maligned and however much we may have had to criticize it even in this book. But it does have the merit of co-ordinating rather than subordinating people. The market and power do not go well together, and anyone who wished to use his strong position vis-à-vis some buyer or seller to establish a dominating relationship of more than transitory duration would find it difficult to do so unless he could count on government support. As long as there exists a genuine market, economic power will remain precarious, and co-ordination will not easily be transformed into subordination. On the other hand, it is one of the most damning things to be held against collectivism in any shape or form that, with the exception only of the few who hold the power to plan and direct, it presses men inescapably into vertical and personal relations of subordination and so robs them of freedom. If the socialists, incorrigible centrists as they are, demand such an economic order in the name of freedom, they afford a most depressing proof of the aberrations of which man is capable when he is blinded by political passion.

However, the medal has an obverse side which we must not overlook. In the measure in which the number of independent people shrinks and in which the large concern and mass organization become typical of our times, in that same measure the market economy loses some of its advantages over the collectivist economy. The web of human relations is impaired even within the market because of concentration within the market. It is true that as long as the market economy survives it will remain incomparably better than the collectivist economy in this respect. In the market economy there always remains some independence, and there are a thousand ways of escape and protection: change of profession or job, free trade-unions, the rule of law, and many others. And if anyone is fed up with the whole thing, as thousands are today in the oppressive air of their welfare state, then he can emigrate to some place where centrism is still kept within bounds. Nevertheless, this is a problem which is certainly becoming more and more serious.

It is only natural that the people who are caught in the hierarchy of dependence should look for some compensation. They will try to loosen the bond of subordination and to narrow down the area of arbitrariness and chance. Every step in this direction will be counted a blessing. Everything possible should, of course, be done to alleviate the human problem of large organizations and concerns. This scarcely needs stressing today, when expressions like "working climate" and "human relations" are on everybody's lips. A justified claim in this connection is that the subordinates should have a share in the responsibility and a say in the affairs or even in the management of their company or organization so that subordination may be mitigated by elements of co-ordination as far as this is at all possible in a setup based, by nature, on subordination.[8]

Another consequence of the process of concentration reveals the full extent of its danger to society and the economy. Suppose that a hitherto independent plumber goes to work in a factory; subordination will disturb his inner balance, and he will try to re-establish it by tending to vote for a political party which promises to make life hard for the bosses. It will depend on circumstances whether he chooses the Socialist or the Communist party. Generally, he will also join whatever union looks after his trade in order to gain at least moral support. Nowadays, when full employment or overfull employment seem to loosen dependence, he may well feel that the price of permanent inflationary pressure—which, he may or may not realize, has to be paid—is not prohibitive as long as the wage-price spiral has not become so obvious that even the trade-union leaders' sophisms cease to make any impression on our plumber's sound common sense. On the other hand, we have seen earlier that these trade-union leaders have their own particular reasons for wanting to press an expansionary wage policy at all costs.

It is understandable enough that the trade-unions should have developed as a kind of defensive reaction to the fact that subordination has become the dominant principle in human relations. This situation has resulted from concentration and the increase in dependent wage earners, and the reaction is mainly moral. But a new danger threatens the dependent worker. The trade-union itself becomes one of those "organizations" which are an expression of growing concentration; it creates, in its turn, new vertical dependences and new hierarchies with an above and a below, with bosses and subordinates. This kind of dependence may become intolerable and overshadow anything that an industrial company may impose on its workers and employees, whenever trade-unions obtain the right to dictate that no one may be employed in a company or profession without belonging to the union. In Anglo-Saxon countries, this occupational monopoly is known as the union shop or the closed shop.

If the courts and the legislature are weak or injudicious enough to tolerate such a monopoly, they must take their share of the blame for a tyranny that is more brutal than any other because it can impose its will by the threat of robbing a man of any chance to earn his living. Both the British press and the American press have recorded hair-raising examples of such trade-union tyranny. Things have come to such a pass that in the United States the "right to work" has assumed an entirely new meaning and now signifies that the worker's job is to be protected from the monopoly powers of the trade-unions.[9] It is an eloquent commentary on our times that even this elementary measure is meeting with such obstinate resistance on the part of the "progressives" and that it has thus far been enacted in only nineteen of America's fifty states.

Whichever way we look at things and whatever consequences we consider, there can be no doubt that if dependent labor, which already is in an overwhelming majority in most industrial countries of the West, goes on increasing, this will create a very disturbing problem. The immense danger of the process lies in its being a process of concentration corresponding to the concentration of firms. The number of firms which transform previously independent workers into dependent workers grows, and so does the average size of firms, for a variety of reasons. At the same time, trade-union power and all of its familiar consequences are strengthened. If we take all this into account, there is every justification for asking the anxious question of whether genuine democracy and a free market economy are, in the long run, compatible with a state of affairs in which the crushing majority of the population consists of dependent wage and salary earners.

The least that can be said about it, and surely something which no one will deny, is that it is a problem whose long-term importance

240

is second to none. It is a key problem which must be solved if democracy and the market economy are to survive. We should not despair of finding a solution, but we must not expect it to be easy and simple. For this reason we cannot discuss it here as it deserves to be discussed. But we can mention three pointers towards a solution. First, we should do everything we can to brake or even reverse the process of dwindling independence whenever and wherever this is possible without real damage to economic rationality. Secondly, we should do everything we can to mitigate the rigidity of vertical subordination as much as the structure of productive organization and the nature of the market economy permit. Thirdly, we should do everything we can to strengthen the counterweights in fields other than labor dependence, the most important of these counterweights being private property.

This program should rally all supporters of our free economic and social order. But the first point may perhaps need some amplification. It implies that if one subscribes to the view that the process which transforms our society into one of dependent labor is disastrous, then one has to face the question how to counteract a further concentration of firms. This is the test case for the decentrist. It is of no avail to look to the government for new compulsion and new legislation, which would only acerbate centrism elsewhere. The decentrist must prove his worth by his support for all the forces, whatever they be, which counteract concentration. Painstaking research would be needed to discover how, ultimately, the government itself, by means of its laws, its tax system, and its economic and social policies, continuously and injudiciously weights the scales in favor of industrial concentration and makes things difficult for small and medium firms and all others who aspire to independence. This has nothing to do with the frequently overrated technical and organizational advantages of scale. The result of such research might be surprising. We might find that a few well-aimed measures, such as the reform of purchase or sales taxes, the aboli-

241

tion, or at least considerable reduction, of double taxation on distributed profits, the radical revision of company law, etc., might be extraordinarily effective in strengthening the position of small and medium firms.[10]

International Centrism

To adopt the program of decentrism has many implications. Once one has done so, it will not do to close one's eyes to the immense problem of the ever growing size of firms and of economic concentration. This problem is fraught with immeasurable dangers for a free society and economy, and the task of countering this danger with all the means appropriate to the decentrist ideal must be tackled. Yet a depressingly large number of decentrists are blind to the problem and give their blessing to the gigantesque in industry. Hardly fewer are those who thoughtlessly join up with centrism when, having done its damage on the national plane, it proceeds to the promising field of international relations.

Under the false colors of international unity, a whole apparatus of international concentration, conglomeration, uniformity, and economic planning has grown up, both within the United Nations and its specialized agencies and on a regional pattern, such as the European Coal and Steel Community. These institutions are waxing in power and provide an ever growing officialdom with privileges, influence, and tax-free incomes. Apart from a few praiseworthy exceptions, the usefulness of this international centralization is fantastically out of proportion to its cost, not to mention the undoubted damage it does. Few are clear-sighted enough to detect the reality behind the semblance of high ideals, and fewer still are courageous enough to speak out—and if they do, they must face a veritable conspiracy of all the *bien-pensants*.

Only a diminishing minority sees that this is centralization of a particularly insidious and dangerous kind, and since the international bureaucracy disposes of powerful means to influence public opinion, even this minority finds it difficult to gain a hearing. International organization goes by many an attractive name, such as "Europe," "supranational sovereignty," "international harmonization," or "fight against Communism," and its worst feature is that it threatens to do away with the last sound remnants of national decentralization and international variety. The shining peak in the distance is the international welfare state, our views on which have, it is hoped, already been made sufficiently clear.

The latest stage in this development is the so-called European Common Market, while the further project for a free-trade area is somewhat less afflicted with centrist features. The economist has reason to be very critical of this project, but this is not our primary concern here.[11] In the present context, the decisive argument is that this project implies a considerable amount of international economic planning and the prospect of more and more concentration and organization in the European economy and is therefore bound to provide a new and powerful stimulus to international centrism. The dependence of the individual and of smaller groups upon large centers will grow enormously, the human and the personal will have fewer chances than ever—and all this in the name of Europe and the European tradition, which owes so much to freedom, variety, and personality.

The danger was lurking in all of the many projects and discussions and records of European economic integration, and today it faces us immediately: it is the danger of *economocracy* finally transferred from the national level to the international level. It means the yet stronger and more inescapable domination of the planners, statisticians, and econometricians, the centralizing power of an international planning bureaucracy, international economic intervention, and all the rest of it. Some few countries of Europe have thus far been able to hold the spirit of Saint-Simonism at bay within their own frontiers, but now it will invade even these, from above, in the form of an European Saint-Simonism, true to the vision of the patriarch of economic planning.

I say this as a man who loves all that the word Europe implies in highest values and loves it with a feeling best described as European patriotism. I am second to none in my love of Europe, least of all to those who have made that word the slogan of officious meddling. To me, it is self-evident that our continent must consolidate if it is to overcome its weakness and safeguard its heritage in the face of threatening dangers; only then can Europe regain her due place in world politics, not solely in opposition to the common foe of the entire free world, but also within the great defensive front of the West and in coming to terms with the colored peoples. But then I also hold the apparently old-fashioned view that this purpose cannot be achieved by the cheapest possible production of automobiles and radio sets but primarily by our continent's regaining its self-confidence, reviving its political and military power, and bethinking itself of the spirit and great heritage in the joint safekeeping of all Europeans.

We can be loyal to Europe only if we preserve her spirit and heritage. The political and economic consolidation of Europe must therefore be such as to embody this loyalty by preserving what is of the essence of Europe: unity in diversity, freedom in solidarity, respect for the human personality and for distinctions and particularities. No matter how far definitions may have diverged in other respects, there has always been unanimity on this capital point: in antiquity, Strabo spoke of the "many shapes" of Europe; St. Stephen of Hungary, in his impressive *Monita* to his heir, warned him that "unius linguæ uniusque moris regnum imbecille et fragile est"; Montesquieu would speak of Europe as a "nation de nations"; and in our own time Christopher Dawson has stressed Europe's character of a "society of peoples." *Decentrism is of the essence of the spirit of Europe.* To try to organize Europe centrally, to subject the Continent to a bureaucracy of economic planning, and to weld it into a block would be nothing less than a betrayal of Europe and the European patrimony. The betrayal would be the more perfidious for being perpetrated in the name of Europe and by an outrageous misuse of that name. We would be destroying what we ought to defend, what endears Europe to us and makes her indispensable to the whole free world.

It is an ominous sign that there should be any need even to argue about the fact that a certain method of European economic integration should be excluded because it is un-European, centrist, and illiberal in the broadest sense of European libertarian thought. Economic nationalism and planning on the continental scale is no progress whatever in relation to economic nationalism and planning on the national scale. Indeed, it is much worse because these tendencies would have much freer scope on the larger territory of a whole continent. If this is agreed, then it should also be clear that there are certain directions in which we should not advance, even by a few steps.

Respect for distinctions and particularities, for diversity and for the small units of life and civilization, and, at the same time, rejection of any form of mechanistic centralization—these are the general principles whose observance alone identifies us as true Europeans who take the meaning of Europe seriously. If we are of one opinion on this, then we also ought to share a certain apprehension aroused by many a misdirected excess of zeal. We should be apprehensive about the activities of the economocrats and technocrats who are busy drawing the blueprints of Europe and creating a giant European organization, all in the name of technical progress. We should be apprehensive, too, about the strange ambition of making Europe a melting pot of nations and civilizations while at the same time treating with contempt precisely that which unifies European civilization at the highest level, namely, the classical and Christian spiritual heritage. We should be apprehensive, finally, about the idea of an European industrialism, which drowns in sheer quantity everything that is qualitative, diverse, varied, immeasurable, and individual and which measures progress in terms of tons of steel, kilowatts, record speeds, and the length of airport runways.

245

Do we want to take as our ideal in Europe mass production and mass cities, as an ideal, moreover, which must not even be challenged? Is it an indisputable advantage for Europe, too, to follow the road of growing concentration and rationalization? Do we not have every reason to fear for all the things which may then be trampled underfoot? Can there be anyone who does not shudder at the thought of an European Detroit disgorging automobiles in such enormous numbers that the density of American traffic is reproduced on our small continent, crowded into a narrow and densely populated space? With men thinking as they do today, all of these are no doubt heretical questions, but they need to be asked all the more insistently as there are only a few who have the courage to pronounce them, for fear of being descried as old-fashioned. This kind of question ought not to be suppressed if we want to bear true witness to decentrism, which, properly understood, is the true philosophy of Europe.

Reckoning without Man

Alarming numbers of people today are prepared to yield without resistance to the centrist trend of the time or even to think that they are doing something highly commendable by promoting it as best they can. There are deep reasons for this, and they are of a spiritual nature. The same trend determines our social philosophy; we think in aggregate, mechanistic, centrist terms and are alienated from man in his concrete individuality. It is not surprising that the social sciences themselves, economics and sociology, increasingly turn to thinking in aggregate and mechanistic terms and to advocating centrism in practical policy. Ortega y Gasset wrote a famous essay on the expulsion of man from art; today we might well add a study on the expulsion of man from economics. Just as, in modern art, man is sacrificed to formless abstraction because he has in reality lost his features and dignity, so do certain theories of the social sciences dehumanize practical policy.[12]

246

In deploring the centrist and mechanistic tendencies in contemporary economics, we revert to a criticism made earlier in this book. We have in mind principally a school of thought which is indissolubly linked with the name J. M. Keynes. It has the significant name of "macro-economics": the economic process is treated as an objective and mechanical movement of aggregate quantities, a movement capable of being quantitatively determined and eventually predicted by appropriate mathematical and statistical methods. The economy takes on the appearance of a giant pumping engine, and it is quite consistent that the science which treats of the economy is turning itself into a sort of engineering science. Equations proliferate, while the theory of prices all but falls into oblivion. Yet the theory of prices, we recall, is the real harvest of a century and a half of economic thought.

A number of other questionable tendencies are connected with this. Excessive specialization furthers the disintegration of the social sciences' body of knowledge; esoteric exposition, taking obvious pride in the handling of mathematics, tends to close off hermetically the separate fields of knowledge; certain intellectual acrobatics, lacking all sense of proportion, tend to lose themselves in hair-splitting arguments and in the construction of "models" without even a basis of approximation to reality; arrogant intolerance is spreading. When one tries to read an economic journal nowadays, often enough one wonders whether one has not inadvertently picked up a journal of chemistry or hydraulics.

It is high time that we should think soberly and critically about these things. Economics is no natural science; it is a moral science and as such has to do with man as a spiritual and moral being. On the other hand, economics does occupy a special position, in so far as its subject, the market economy, objectivizes subjective matters to such an extent that we can borrow methods from the natural sciences. This special position confers upon economics all the opportunities and charms of a "borderline science"—but also all the dangers.[13] We can use mathematics for illustrating and precisely formulating functional relationships between quantities, and few contemporary economists wholly condemn such use. But this method has its dangers. Unless its user is very careful, he may be tempted into pushing the critical borderline territory—the territory between the human and the mechanical—too far into the realm of mechanics, statistics, and mathematics, and he may neglect what is left this side of the frontier, namely, the unmathematically human, spiritual, moral, and, for this reason, decidedly unquantifiable. We would be wise to use the technical methods of the natural sciences only occasionally and for purposes of illustration; the possible gain is disproportionately small in comparison with the effort and the dangers involved. *Parturiunt montes nascetur ridiculus mus*—this would be a fitting motto for many a study of this kind.

It is a serious misunderstanding to wish to defend the mathematical method with the argument that economics has to do with quantities. That is true, but it is true also of strategy, and yet battles are not mathematical problems to be entrusted to an electronic computer. The crucial things in economics are about as mathematically intractable as a love letter or a Christmas celebration. They reside in moral and spiritual forces, psychological reactions, opinions which are beyond the reach of curves and equations. What matters ultimately in economics is incalculable and unpredictable. No more must be expected of the mathematical method than it can, at best, perform. It would be difficult to name any sound economic theory which could be discovered only by this method or, indeed, any single one which was, in fact, so discovered. There are profound reasons for this, for every economic theorem which can be demonstrated only by means of mathematics and is not evident without them deserves the greatest mistrust. In the face of such attempts, the best answer is a phrase which one of the economists of the old Vienna school used to be fond of in cases of this kind: "Rather than be surprised, I prefer not to believe it."

Voltaire's remark, which Goethe once approvingly quoted in a letter to Zelter, holds here as much as anywhere: *"J'ai toujours*

248

remarqué que la géométrie laisse l'esprit où elle le trouve." Only too often does mathematical economics resemble the children's game of hiding Easter eggs, great jubilation breaking out when the eggs are found precisely where they were hidden—a witty simile which we owe to the contemporary economist L. Albert Hahn. The same irreverence, I am afraid, is due mathematical economics when it pretends to furnish us precise results. In a science in which the subject matter simply precludes the exactness of mathematics and the natural sciences, such a claim is bound to raise the gravest misgivings. We reply that it is better to be imprecisely right than to be precisely wrong.[14]

After the First World War, a French statesman said: *"Un homme qui meurt—ça m'émeut. Quinze cent mille hommes—c'est de la statistique."* This is as true as it is bitter. Economics should not overlook the lesson. We do, of course, need a sort of technical shorthand language in our science. We speak of supply and demand, the purchasing power of money, volume of output, volume of saving, volume of investment, not to mention the hog sector, and we cannot go on repeating every time that behind these pseudo-mechanical aggregates there are individual people, with their thoughts, feelings, value judgments, collective swings of opinion, and decisions. But we ourselves should not forget it, and we should not play with aggregates as with building blocks.

Ingenious tools of analysis have been devised for the examination of the economic process, and some of them we would not want to do without. But in handling such concepts as the "elasticity" of supply and demand, the "multiplier," the "accelerator," and so on, we should always remain conscious of the narrow limits of their fruitful and harmless application. They simulate a scientific and mathematical precision which does not really exist. They are not physical constants like the acceleration due to gravity but relations dependent upon the unpredictable behavior of men.

Let us illustrate the point with an example. I recall a rather pathetic study, published many years ago by General Motors, which

contained the results of years of research, in an econometric labora-
tory especially founded for that purpose, about the behavior of the
demand for automobiles. The results were disappointing. At the
end of their labors the authors had to confess that, notwithstanding
a lot of mathematical symbols and figures, they were no wiser than
before. It had been worked out how buyers had behaved in the
past, but how they would behave in the future was as uncertain
as ever. The only bright feature was that the authors frankly admitted the crying disproportion between effort and result. All that
had been proved—although no proof was needed—was that while
it may be interesting and even useful to discover the coefficients of
the elasticity of demand for any product, such figures ultimately
have only historical significance. This method is one way of research into economic history; it is an instructive method and one
which facilitates the assessment of future tendencies, but this assessment always comes up against the basic uncertainty and unpredictability of the future in economic life. Any extrapolation of past facts
is misuse and rests on a misconception. All the unforeseeable forces
which move human history as a whole may at any moment modify
supply and demand in a manner which defies econometric treatment and may continually create new and unexpected situations.

A conception which reduces the economic process to a functional
relationship of aggregates susceptible of being defined in terms of
mechanics and calculated by mathematical methods is, however,
inevitably bound to end up in the claim that these same methods
permit forecasts which are more than the mere weighing of probabilities. This claim is quite obviously unjustified. The chain of
humiliating defeats which econometric prophecies have suffered in
the course of decades is therefore not surprising. What is surprising is the refusal of the defeated to admit defeat and to learn greater
modesty.

Is it necessary to cite examples? A few months before the beginning of the greatest economic crisis in history, in the spring of
1929, the most distinguished American economists were talking

about the happily secure equilibrium of an economy running in top gear. Where are the prophets of the decline in the birth rate? Population statistics seem a firm enough ground, yet these theorists had calculated not so long ago that a rapid decline in population growth was absolutely certain and had gone on to make precise recommendations for economic and social policy. Where are the economists who had abused the Keynesian theory—not least because of the mistaken forecasts of a declining birth rate—in order to deny that our economic system had any natural growth potential? They have left behind only the theory of the "mature economy," an intellectual fossil of the Great Depression, that same Great Depression whose threatening symptoms had been misinterpreted so shortly before its outbreak. And consider the damage done by the unfortunately all too influential prophets who were Keynes's obedient disciples. They forecast another depression to follow the Second World War and put their money on the wrong horse once more. They warned governments of being too optimistic about the peace and advised them to counteract the coming depression by doing everything in their power to increase purchasing power, with the result that most governments, in fact, pursued a wholly untimely and inflationary policy of full employment. Must we still recall the prophecies of gloom with which the adversaries of the German market economy accompanied its revival and were proved wrong again and again? And what about the Swedish economists who, in 1948, predicted another American depression, which again failed to materialize, and thereby misled government and central bank into plunging a rich and prosperous economy into a disorder which it has not been possible to straighten out to this day?

It would be interesting to know how certain forecasters come to terms with contrary facts. There are those, for example, who, for more than ten years, have spoken of the "permanent dollar gap" and "structural balance-of-payments deficits," as if these were objectively determined long-period phenomena; but meanwhile, some of the structurally weakest countries, the losers in a war which

251

devastated them, Germany and Austria (and Japan, as long as it pursued anti-inflationary policies), have resoundingly disproved these self-assured predictions and have re-established a balance-of-payments equilibrium ahead of all other countries, including the richest. Of what avail is the vast effort of mathematical speculation about import and export elasticities, terms of trade, and other props of these theories when such countries are so tactless as to prove empirically that the classical balance-of-payments theory is right?[15]

In recalling these painful memories of disproved prophecies, we do not, naturally, mean to imply that it is not a legitimate task to evaluate the effect of present trends on the future and to weigh probabilities against each other. Nothing is more natural or more necessary than this. But we should have learned to do it with a greater mistrust of deceptive mathematical and statistical calculations and alleged data—including the psychological ones with which Keynes operated. We should instead base our calculations on man, not on some fictitious man who fits into our equations, but man as he is, with his oscillation between hope and fear, with his whims and passions, with his susceptibility to mass opinions and mass tempers, his fluctuation between quiet contentment and novelty-craving boredom, his dependence upon others and upon facts, and with the imperfection of his knowledge of these "data." If we want to guard against errors and disappointments in the analysis of economic processes and still more in our guesses about the future, then we must bear in mind that the science which treats of these processes, economics, is a science of the behavior of man in a certain sphere and in certain circumstances. What Epictetus said of social affairs is as true as ever, namely, that the decisive matters are not facts but men's opinion about facts or even opinions about opinions, even though they are, of course, linked to the facts.

We now see also why one particular form of mechanistic and centrist theory underlying economic policies deserves a very skeptical reception. I have in mind so-called national budgeting, which is a statistical exposition of the interrelationships of aggregate quantities (income, consumption, saving, investment, government receipts and expenditure, foreign trade, etc.) over a given period of time and which is intended to serve as an indication for future economic policy. In so far as this is no more than a kind of statistical economic history, setting out the estimated development of these magnitudes during the past year, there is, of course, no objection. But the very violence with which the uses of national budgeting are defended should put us on our guard: even when it is not explicitly stated, these exercises have an ulterior purpose beyond the mere satisfaction of our historical and statistical curiosity.

We are once more in the presence of economocratic aspirations to power. The obvious intention is to turn national budgeting into a tool for mastery over the circular flow of the economy, and to this end the mathematical and statistical "exactness" of the results is invoked and the claim put forward that future developments can also be calculated. A warning is in order, however. Either these calculations are just another form of leaving man out of the accounts, that is, they are mere guesses which come to grief on the eternal uncertainties and therefore constitute a danger for any economic policy based thereon, or they are a permanent temptation to use planning and intervention in order to force the recalcitrant facts into the predicted pattern. In both cases, no good can come from national budgeting. Moreover, there is no clearly established, visible boundary between using national budgeting as a mere—even though possibly mistaken—indication for economic policy and using it for purposes of economic planning.

The enthusiasm for this fashionable product of mechanistic and centrist thought should, in any event, be damped by the sobering reflection that the countries with the soundest economic and currency policies are those which do not go in for national accounting, while the countries in which this method is at its apogee have sickly economies and unusually strong inflationary pressure (the Scandinavian countries, the Netherlands, Great Britain, France). The most charitable interpretation of this state of affairs would be that the former countries were able to do without national accounting and still make the right decisions in economic and monetary policy, while the latter were not deflected from the wrong decisions even by national accounting. In fact, it is likely that national budgeting, as an essential instrument of economocracy, is more than a little responsible for the errors.

The failures of the mechanistic and centrist approach in economic forecasting are so numerous and blatant that it is astonishing that the underlying theory seems to digest these failures without losing prestige. It is even more astonishing that the protagonists of this approach are so utterly unrepentant. This is a problem which will repay some reflection. Is the British economist Professor Ely Devons right when he says (*Lloyds Bank Review* [July, 1954]) that the role of statistics in our societies has a striking resemblance with some of the functions of magic and divination in primitive societies? "Statistical magic, like its primitive counterpart, is a mystery to the public; and like primitive magic it can never be proved wrong. . . . The oracle is never wrong; a mistake merely reinforces the belief in magic. It merely demonstrates conclusively that unless you do everything the right way you will get the wrong answer. So with us, bad forecasts rarely discredit statistical magic; they merely serve to demonstrate that the basic figures were bad, that the model was wrong or the statistician mistaken in his interpretation. . . . Next time we shall use better figures, better models, and of course the statisticians and econometricians today would never make the silly misinterpretations made in 1944, 1945 or 1946. We are convinced, rightly or wrongly, that this is the scientific procedure and we are going to stick to it." These bitter words reveal the full measure of the disappointment in which statistical and econometric research has ended in England. We would do well to take to heart the truth they contain. But what seems even more important to me is that the true reason why this method triumphantly survives all disappointment is that it is an indispensable instrument of economocracy.

We know that the method is wrong, and, being decentrists, we are convinced that it serves a bad purpose. The purpose is bad because it is to centralize and overorganize the economy and society in a way which neglects the human element. If we manage to get away from this mechanistic and collectivist way of thinking, we shall, among other things, also see the position and function of the entrepreneur in their true light. The essentials about this can be expressed in one sentence: The entrepreneur is like a ship's captain whose principal task is continuous navigation on the sea of the market, which is unpredictable because it depends on human nature. His function, which is perennial and indispensable for the course of economic life, is to balance supply and demand and continually to adjust production to changing consumption. In discharging this function, he always has to reckon with the uncertainty and unpredictability of the market factors, and his success depends on the extent to which he achieves this adjustment in spite of extraordinary difficulties. A man is an entrepreneur to the extent that he has learned to weigh probabilities against each other and to organize production and sales accordingly; and he is a successful entrepreneur exactly to the extent that he gets the better of the uncertainty of future market situations.

To think of the entrepreneur as a navigator should make a number of things clear to us. If a firm is to be successful and, in the degree of its success, to discharge its economic functions in society, it must primarily be orientated towards the market and must continually battle with its uncertainty and unpredictability. The firm's face is turned outwards, towards the market, and competition among firms is a continuous struggle to gain a start on the others in the matter of knowing or correctly assessing or discovering new opportunities for production or sales. The firm which so gains a start obtains, for the latter's frequently very brief duration, a special, privileged position which could easily be mistaken as monopolistic if it were not at once threatened and soon removed by those hard on the heels of the leader in the field. But the possibility of temporary advantage, the dynamic character of competition, is indispensable in order to spur the entrepreneur towards the best possible fulfillment of his function. As such, it is an essential part of any genuine market economy.[16]

The same interpretation holds if we include the many cases in which the entrepreneur does not take the market as given but influences it, or even creates it or opens it up. But no entrepreneur can influence, create, or open up a market unless there is harmony between what he, the pioneer, innovator, organizer of production, or bringer of new wares, wants to do and the latent desires and reactions of the market, that is, of the people who accept one thing and reject another. It is always the market, with all of its unknowns, which decides and which ratifies or invalidates the entrepreneur's plans. However much the entrepreneur would like to shake off or reverse this relationship, he remains the servant of the market. His compliance is rewarded and his disobedience punished as long as the market is genuine, that is, ruled by competition. Even if, like Antonio, he be a "royal merchant," he can claim this title of honor only if he is also "the greatest servant of the commonwealth," that is, the market. He can be an absolute king only to the extent that monopoly enables him to forget his subordinate position.

The entrepreneur's function of navigating unpredictable seas is also the origin and justification of profit, as it is conceived in pure theory. This has always been the most disputed and least understood form of income, largely because one usually mentions only the entrepreneur's profit, but not its twin, his loss. Since Frank H. Knight's now classical work *Risk, Profit, and Uncertainty* (1921), there ought to remain no doubt about the fact that profit and loss are linked to a basic uncertainty of the future pattern of supply and demand and that they correspond to the entrepreneur's success in assessing probabilities, as it is incumbent upon him to do, and in quickly adapting himself to a changing market. In a dream world of perfect knowledge, there could be no difference between sales price and production cost (in the widest sense) and hence no profit or loss.[17]

Now there is obviously a profound and positive meaning in the fact that success in groping among uncertainties is so promptly rewarded or lack of success equally promptly punished. This specific entrepreneurial activity is not only immensely important but also extraordinarily difficult. Only in the pathological situation of an inflationary sellers' market is it degraded to a sort of amateur sport; otherwise it remains navigation which requires the experience, resourcefulness, and sound intuition of the master mariner. Neither textbooks nor training courses, neither statistics nor electronic computers, can replace these qualities. This is why we need the entrepreneur and at the same time an economic order such that the best selection of these captains of the market according to their qualifications is always ensured and such that there is always an effective inducement towards maximum performance and careful decisions on the part of the entrepreneur. Profit and loss (and, ultimately, bankruptcy) together provide this inducement, and it should be one of our major preoccupations to see to it that it does not lose its force, even in the modern large company, with all its legal and organizational complications.

As consumers, we all benefit from this arrangement. Yet strangely enough, we can be persuaded only with difficulty, and not always successfully, to recognize it. This is one of the things which has never been satisfactorily explained. We often behave as if the whole arrangement had been invented solely for the pleasure and the advantage of entrepreneurs, and therefore we tend to regard them as the natural advocates of the free market economy. This really is strange. It would appear natural, on the contrary, that they should consider it inconvenient and tiresome to be saddled with this inducement system of the market. At any rate, many entrepreneurs display such a feeling in their attempts to withdraw from their position as navigators on the open and uncertain sea of the market. But if we abolish the market and replace it with planning authorities, or even if we allow government intervention and monopoly to restrict the market and turn it into a stagnant pool, then we do not need the entrepreneur. If the entrepreneur, like a seasick captain, wants to escape the whims and risks of the market and to take refuge in the safe port of planning or in security guaranteed by government or cartels, he makes himself redundant.

As we said, this may be quite natural and human. The motives for such an attitude may be respectable, but they are not precisely glorious. In any case they testify to a regrettable short-sightedness. On the other hand, there is another form of discontent among entrepreneurs which deserves our full respect, even though it may be unfounded. Entrepreneurs protest when economic theory seems to attribute to them the role of mere automata which achieve the common good by simply fulfilling the function which competition assigns them, calculating their advantage without thought of any higher ethical purposes. With some slight exaggeration, this attitude may be described rather like this: Let the entrepreneur be content to produce shoes as cheaply and as well as possible and to pay the factors of production, especially labor, according to their market value; if he also happens to be a decent chap (or is clever enough to be one, as the new theory of "human relations" advises him), he can be more than satisfied.

If entrepreneurs protest against such a moral and intellectual depletion of their existence, this is not only natural, but honorable and encouraging. No man can live a full life by credit and debit entries alone, even though the honest fulfillment of a responsible task is in itself no mean thing. But it is a misunderstanding to believe that our economic order expects such emptiness of the entrepreneur. We hardly need to dwell, in the final pages of this book, on the nature of this misunderstanding and the reasons for it. We want to stress the point, however, that one can take such an empty view of the entrepreneur's activities only by starting out from the wrong concept of the economy as a mechanical process. The dehumanization of theoretical economics necessarily includes a human devaluation of the entrepreneur, as of all other economic groups. As against the physics of the economy, we have to underscore its psychology, ethics, intelligence—in short, its human elements.

The human elements in the economy also enter into the argument in a very specific sense which bars the way to all centrist and mechanistic conceptions and reveals once more, in a prosaic manner, how wrong we can go in reckoning without man. Let us leave the lofty peaks of philosophy and take the low road of sober common sense. We have already dealt with the question of what centralization means for the highest human values, for man's soul, freedom, community, and ultimate destiny. Now we ask simply: Where are the people, where, above all, are the leaders, who can take on and bear the burden of centralization? Does not centralization come up against sheer physical limits, limits which have become quite obvious by now and which make the principle of centralization not only reprehensible but downright impracticable?

Here we meet the centrists on their own plane of the tangible and practical, from which they are wont to look down upon us as dreamers romantically worrying about the fate of mankind. We, the dreamers and romantics, are unimpressed by super-organization, centralization, Gargantuan concerns, machine giants, mammoth towns, and titanic plans. Undaunted, we keep quoting Montaigne's wise words—that even on the highest stilts we still run with our own legs and even on the highest throne sit with our own rumps. We humbly ask how all of these much-vaunted things are going to be done in the absence of a sufficient number of people with the required mental and physical qualities. The claims on the human mind, heart, stomach, and liver are simply more than most men can meet, since their spiritual and physical resources go only so far and no farther. Here is the weakest link in the whole chain, and we cannot but accept this weakness with humility and modesty. Neuroses, heart diseases, and stomach ulcers are the final irrefutable arguments against centrism of every kind. To disregard them is wanton *hubris*, but we may not be far wrong in interpreting our times as a concatenation of *hubris* and nemesis.

It is time to return to a thought which we touched upon in the first chapter. We spoke of the excess of government intervention which vitiates the market economy, even in its model countries. We pointed to a number of tangible and measurable consequences of this hodge-podge system but left the essential thing unsaid. The essential thing, as always, lies in the realm of the immeasurable and imponderable, and this multiplies the danger in an age such as ours, which has lost the feeling for these to such an alarming extent. It is easy enough to say that, after all, we have come to quite tolerable terms with this regrettable excess of government intervention, that the Germans, or even the Norwegians and the British, are reasonably well off, that they can buy anything that may be their—not unduly immodest—heart's desire. It sounds soothing, too, if we stress the extraordinary resilience of the market economy in adjusting itself to government intervention and overcoming the resulting difficulties, its robustness, stamina, and ostrich stomach. But we know how illusory all of this is.

Let us leave aside what we have already said earlier in Chapter I. But quite apart from that, who can measure the sum of nervous tension, the time and effort wasted on the double-front war against the market and the authorities, the nights spent poring over forms, the negotiations, useless trips, the irritation and vexation due to overbearing authorities? Competition in the market economy is wearing enough, but when it is, in addition, necessary to battle constantly with officials, to take account of their commands or prohibitions, to worry how to steer the firm not only through the whirlpools of the market but between the artificially created cliffs of government intervention and financial policy—how long can anyone stand this double strain? We are all ordinary men with limited strength. The much-vaunted resilience of the market economy is, in the last resort, the resilience of the people on whom rests the responsibility; the robustness of the market economy is that of the bodies and nerves of all those who have to carry the double burden of market and officialdom; the ostrich stomach of the market economy means nothing else but the stomach, heart, and other organs of the victims of this overcentralized and overorganized system. It is in the consulting rooms of heart, stomach, and nerve specialists that the balance sheet of the system has to be drawn up.

Who can measure the sum of happiness, contentment, well-being, sense of fulfillment, and elementary freedom which is destroyed every day and every hour? The more we adulterate the market economy with admixtures of intervention, the higher rises the watermark of compulsion, the narrower becomes the area of freedom. What distinguishes the centrist from the decentrist is that the former makes so much lighter of the growth of compulsion than the latter.

Our world suffers from the fatal disease of concentration, and those—the politicians, leading personalities of the economy, chief editors, and others—in whose hands the threads converge have a task which simply exceeds human nature. The constant strain is propagated through all other levels, down to the harassed foreman and his like. It is the curse of our age. It is a curse twice over because these men, who can do their duty only at the peril of angina pectoris, lack the time for calm reflection or the quiet reading of a book. This creates the utmost danger for cultural leadership. Who can imagine nowadays an age like that of the Younger Pitt, when, as he relates, the Chancellor of the Exchequer in Great Britain needed no private secretary because the extent of business did not justify one? Or who can imagine the way of life of Alexander von Humboldt, who could deal by himself with his annual correspondence of about three thousand letters and still remained one of the foremost thinkers of his generation and reached the age of nearly ninety years?[18]

We shall not inquire about the simple happiness which is at stake. All that is self-evident. We only have to say this: our centrist civilization, which has become more and more remote from man and the human scale, has reached the point where its own continued existence is at stake.

Notes

CHAPTER I

1. What Eric Voegelin means by this apt expression is explained in his paper "The Origins of Totalitarianism," *The Review of Politics* (January, 1953): "The Christian faith in transcendental perfection through the grace of God has been converted—and perverted—into the idea of immanent perfection through an act of man." The idea and all of its consequences were fully developed by Voegelin in his book *The New Science of Politics* (Chicago, 1952).

2. Ricarda Huch, *Untergang des Römischen Reiches Deutscher Nation* (Zürich, 1954), 218f. If, on the other hand, we want to know what goes on in the soul of the modern city dweller and mass man, we have but to consult such books as the famous Kinsey Report or an enlightening British book by B. Seebohm Rowntree and G. R. Lavers entitled *English Life and Leisure: A Social Study* (London, 1951). Cf. Russell Kirk, *A Program for Conservatives* (Chicago, 1954), 101ff., and *idem, Beyond the Dreams of Avarice* (Chicago, 1956), 187ff.

3. I have discussed this in more detail in my book *L'économie mondiale aux XIXe et XXe siècles* (Geneva, 1959), 10ff.

4. Alfred Weber (*Farewell to European History* [London, 1947]; *Kulturgeschichte als Kultursoziologie* [2nd ed., Munich, 1950]) thought it very likely that a new human type might emerge and described this "fourth man" in a bitterly disillusioned

manner. Cf. the critical comment in my Introduction to the collection of essays *Kommt der vierte Mensch?* (Zürich, 1952).

5. See the works of poet-philosophers Max Picard (*The Flight from God* [London and Chicago, 1951]; *Hitler in Ourselves* [Chicago, 1947]; *World of Silence* [London and Chicago, 1952]; *Zerstörte und unzerstörbare Welt* [Erlenbach-Zürich, 1951]) and Rudolf Kassner (*Das neunzehnte Jahrhundert* [Erlenbach-Zürich, 1947]).

6. I have in mind especially F. A. Hayek's *The Road to Serfdom* (London, 1944). For an account of what the world situation looked like to many people fifteen years ago, see, for example, E. H. Carr's *Conditions of Peace* (London, 1942) and my review of the German edition of this book in *Neue Schweizer Rundschau* (October, 1943).

7. See my book *The Solution of the German Problem* (New York, 1947), especially the final chapter, which diagnoses the economic ailment and prescribes a cure that is more or less the same as the one subsequently adopted. See also my contribution "Das deutsche Wirtschaftsexperiment—Beispiel und Lehre" to A. Hunold (ed.), *Vollbeschäftigung, Inflation und Planwirtschaft* (Erlenbach-Zürich, 1951), as well as my *Ein Jahrzehnt sozialer Marktwirtschaft in Deutschland und seine Lehren* (Köln-Marienburg, 1958).

8. See especially the report on the German economic situation which I prepared in the summer of 1950 at the request of the Adenauer government: *Ist die deutsche Wirtschaftspolitik richtig?* (Stuttgart, 1950.)

9. The argument that a tax ratio (that is, the proportion of national income absorbed by the budget) exceeding, say, one-fourth of the national income must, in the long run, be inflationary has been propounded, above all, by Colin Clark, first in his paper "Public Finance and Changes in the Value of Money," *Economic Journal* (December, 1945). There may be different views about the critical percentage (cf. the discussion in *The Review of Economics and Statistics* [August, 1952]), but the principle itself seems unexceptionable. For an important contribution on this subject, see G. Schmöl-

ders, "Steuersystem und Wettbewerbsordnung," *ORDO, Jahrbuch für die Ordnung von Wirtschaft und Gesellschaft* (1950).

10. Henry C. Simons, "Reflections on Syndicalism," *Journal of Political Economy* (March, 1944), reprinted in *Economic Policy for a Free Society* (Chicago, 1948); Fritz Machlup, *Monopolistic Wage Determination as a Part of the General Problem of Monopoly* (Chamber of Commerce of the United States, 1947); Goetz Briefs, *Zwischen Kapitalismus und Syndikalismus* (Bern, 1952); Hans Willgerodt, "Die Krisis der sozialen Sicherheit und das Lohnproblem," *ORDO, Jahrbuch für die Ordnung von Wirtschaft und Gesellschaft* (1955). See also Chapters IV and V of this volume.

CHAPTER II

1. On the subject of mass society, see Hendrik de Man, *Vermassung und Kulturverfall* (Bern, 1951); Robert A. Nisbet, *The Quest for Community* (New York, 1953); David Riesman, *The Lonely Crowd* (New Haven, 1950); Hans Freyer, *Theorie des gegenwärtigen Zeitalters* (Stuttgart, 1955); Kirk, *A Program for Conservatives;* A. Hunold (ed.), *Masse und Demokratie* (Erlenbach-Zürich, 1957).

2. While the Protestant churches do not take up any uniform position on this point, the Catholic church has always viewed population growth with optimism and sympathy, or at any rate has rejected the possibility of influencing the birth rate. But it has no doubt been sensitive to the conflict between its own position and the undeniable problems of demographic expansion. There are indications that such awareness is growing. A striking example is the positive point of view on birth control which the Jesuit Father Stanislas de Lestapis took up at the World Population Conference in Rome in 1954. A few years earlier, in 1951, Pope Pius XII had sanctioned the practice of observing the "safe periods." (*New York Times,* September 10, 1954)

In Japan, the dissemination of information and the establishment of advisory centers, often under government auspices, have led to a reduction of the birth rate from 34 in 1947 to 20 in 1954. Large private firms, too, are helping to promote birth control (Rev. W. A. Kaschmitter, "Japan's Population Problem," *Migration News* [1956], 1).

3. On the population problem, see Robert C. Cook, *Human Fertility* (London, 1951); Harrison Brown, *The Challenge of Man's Future* (New York, 1954); Edward Hyams, *Soil and Civilization* (London and New York, 1952); Frank L. McDougall, "Food and Population," *International Conciliation* (December, 1952); Henry H. Villard, "Some Notes on Population and Living Levels," *The Review of Economics and Statistics* (May, 1955).

This may be the place to quote the following passage from a German newspaper: "The more Europe becomes industrialized and motorized and the more uniform the pattern of consumption becomes, the more people long to get away from the crowd, at least during their holidays, and to escape into undisturbed calm. The seacoast, for all its being a rather melancholy place sometimes, does offer these people guaranteed solitude, at least on one side, provided they have a house right on the shore." (*Frankfurter Allgemeine Zeitung*, October 13, 1956) No comment is needed.

Collectors of historical detail and anecdotes will be interested to read what Wilhelm Grimm wrote in a letter in 1823: "Much as I should like to be in Berlin for a while, I have misgivings about living there or in any large town. Last year, someone I know in the civil service there told me that all through the summer he didn't get beyond the city gates into the country more than three times." (Wilhelm Grimm in a letter to Savigny dated September 9, 1823, in *Briefe der Gebrüder Grimm an Savigny* [Berlin-Bielefeld, 1953], 331) Jacob Grimm wrote much the same thing about Hamburg in a letter dated November 5, 1817.

4. On this subject, reflect on what Jules Romains has written in *Le Problème numéro un* (Paris, 1947), 71-85. We shall be led to similar considerations in the subsequent chapter on inflation.

5. For more detail, see my lectures on "Economic Order and International Law" at The Hague Academy of International Law

(*Recueil des cours 1954* [Leiden, 1955], 207-70) and my book *Internationale Ordnung—heute* (Erlenbach-Zürich, 1954), 101-63. The problem is stated very clearly by the British historian Herbert Butterfield in *Christianity, Diplomacy and War* (New York, 1953), 79-101.

6. See L. Albert Hahn, *Common Sense Economics* (New York, 1956), 183-84; A. R. Sweezy, "Population Growth and Investment Opportunity," *Quarterly Journal of Economics* (November, 1940); S. Enke, "Speculation on Population Growth and Economic Development," *Quarterly Journal of Economics* (February, 1957).

7. In the German Bundestag recently, according to the record of the proceedings, there occurred an incident in which a three-word Latin phrase (*vigilia pretium libertatis*, NATO's motto) provoked a Social Democratic Deputy to this angry interruption: "Speak German in the German Parliament!" (*Frankfurter Allgemeine Zeitung*, May 26, 1956) The battle which is being waged against the teaching of Latin, even in Italy, must be explained by the same kind of resentment. The American Russell Kirk (*The Conservative Mind* [2nd ed., Chicago, 1954], 381) aptly remarks: "When our universities and colleges devote themselves to turning out specialists and technicians and businessmen, they deprive society of its intellectual aristocracy and, presently, undermine the very social tranquillity upon which modern specialization and technical achievement are founded." We might add that even on the practical plane classical education usually comes off best because the intellectual and moral discipline it imposes, as no other type of education does, provides excellent training in the rapid grasping of any problem, no matter what kind. It is highly characteristic of modern mass society and the concomitant obsession with social affairs (cf. my *Mass und Mitte* [Erlenbach-Zürich, 1950], 60ff.) that the favor denied the humanities is instead lavished upon the social sciences. We can get some idea of how exaggeratedly popular the latter are when we learn that there are forty thousand college graduates in the United States who describe themselves as "social scientists," which means that there is one social scientist for every one hundred farmers or twenty-five school-

teachers or five physicians (William Schlamm, in *Faith and Freedom* [February, 1955]).

The mixture of puerile rubbernecking and abject fear with which the non-Communist world recently reacted to the Communist showpiece of the first artificial earth satellite gives one a horrifying insight into our society's spiritual condition. One of the silliest reactions, though by no means surprising, was that people earnestly proposed that in order to catch up with the Russians' alleged technical lead, we should hurriedly transform all our schools into factories turning out engineers, physicists, and chemists and throw Thucydides, Cicero, Shakespeare, and Goethe on the scrap heap. It is unnecessary to say that great achievements in the fields of physics, chemistry, and mathematics thrive best in the soil of classical education, with its strict training of the mind, that the abandonment of this education would be cultural suicide, and that the Americans would be well advised to retransform their schools into institutions where the mind is disciplined by classical education.

8. John Stuart Mill's testimony can be found in Chapter III of his famous essay *On Liberty* (London, 1859). Several decades later, American conditions inspired Herman Melville, author of *Moby Dick,* to write the following lines:

> Myriads playing pigmy parts—
> Debased into equality:
> In glut of all material arts
> A civic barbarism may be:
> Man disennobled—brutalized
> By popular science—atheized
> Into a smatterer:
> Dead level of rank commonplace:
> An Anglo-Saxon China, see
> May on your vast plains shame the race
> In the Dark Ages of Democracy.

(Quoted from Erik R. von Kuehnelt-Leddihn, *Liberty or Equality: The Challenge of Our Time* [London, 1952], 25.)

9. Cf. my essay "Die Kellerräume unserer Kultur," *Neue Schweizer Rundschau* (November, 1949).

10. Cf. *ibid.*

11. On the delightful subject of the history of children's books and their literary significance, see Paul Hazard, *Books, Children and Men* (Boston, 1947) and J. Dyrenfurth-Graebsch, *Geschichte des deutschen Jugendbuches* (Hamburg, 1951).

12. The decay of our civilized languages can be measured by various indicators, not least by disregard of accepted usage, increasing impoverishment of grammar and expression, coarseness of taste, and lack of logical discipline. Eventually, people cease even to be aware of the downward trend. The phenomenon can be observed in all countries, even in one that is as linguistically sure footed as France.

13. "Meanwhile, I have been passing the time with Niebuhr's and Volney's travels in Syria and Egypt, and I would thoroughly recommend this sort of reading to anyone discouraged by the present bad political outlook. Such books bring home to us what a blessing it is, after all, to have been born in Europe. It is really inexplicable that man's creative forces should be active in only so small a part of the earth, while all those vast peoples simply do not count as far as human progress is concerned." (Friedrich Schiller in a letter to Goethe dated January 26, 1798)

14. Peter Viereck, *Shame and Glory of the Intellectuals* (Boston, 1953). See also the refreshing book by Charles Baudouin entitled *The Myth of Modernity* (London, 1950).

15. De Man, *op. cit.*

16. No one who knew the United States, say, twenty-five years ago and looks at the present situation can help being struck by the horrifying decline in the average level of reading matter as a result of the pressure of mass culture. Most of the then famous and widely read American periodicals, such as *American Mercury, Scribner's, Century, Harper's Magazine,* or *Atlantic Monthly,* have either disappeared or have sunk into insignificance, thereby depriving the public of a lively forum of discussion. Today's scene is dominated

by illustrated mass magazines and synthetic products like *Reader's Digest,* while serious publications have to wage a constant struggle for survival. As regards books, the mere fact that wage increases have accentuated the need for mass production (today a book can hardly be published without loss unless it can sell at least ten thousand copies) makes it increasingly difficult to publish books that do not cater to mass tastes (cf. my essay "A European Looks at American Intellectuals," *The National Review* [November 10, 1956]). To pretend that the trend of developments is any different in European countries would be foolishly self-deluding and smug. The decline of German literature, also chiefly due to the impact of mass culture, has been frankly described by Walter Muschg in his *Die Zerstörung der deutschen Literatur* (Bern, 1956): "The creative writer has lost his place in society because society itself is disintegrating and because its conception of literature is called in question. Anonymous forces, stronger than any individual, are conquering the world. The creative writer's most dangerous enemy is not political dictatorship but the technical-mindedness of the masses, who would rather have a comfortable life than freedom." The scene is dominated by illustrated tabloids of a level which could hardly be lower, shamelessly sensation-mongering magazines have sales figures that approach the American ones, and screaming headlines are considered good journalism. As regards England, there is an impressive though perhaps rather too pessimistic description in Kirk, *Beyond the Dreams of Avarice,* 298-310.

17. The following quotation is from Alexis de Tocqueville's *Democracy in America* (New York, Knopf, 1948), Vol. II, Bk. IV, Ch. 6, p. 318: "I think, then, that the species of oppression by which democratic nations are menaced is unlike anything that ever before existed in the world; our contemporaries will find no prototype of it in their memories. I seek in vain for an expression that will accurately convey the whole of the idea I have formed of it; the old words *despotism* and *tyranny* are inappropriate: the thing itself is new, and since I cannot name, I must attempt to define it.

"I seek to trace the novel features under which despotism may appear in the world. The first thing that strikes the observation is an innumerable multitude of men, all equal and alike, incessantly endeavoring to procure the petty and paltry pleasures with which they glut their lives. Each of them, living apart, is as a stranger to the fate of all the rest; his children and his private friends constitute to him the whole of mankind. As for the rest of his fellow citizens, he is close to them, but he does not see them; he touches them, but he does not feel them; he exists only in himself and for himself alone; and if his kindred still remain to him, he may be said at any rate to have lost his country."

18. Many important contributions to the subject of mass democracy have appeared in recent years: Kuehnelt-Leddihn, *op. cit.*; W. Martini, *Das Ende aller Sicherheit* (Stuttgart, 1954); G. W. Keeton, *The Passing of Parliament* (London, 1952); Nisbet, *op. cit.*; F. A. Hayek, "Entstehung und Verfall des Rechtsstaatsideals," in A. Hunold (ed.), *Wirtschaft ohne Wunder* (Erlenbach-Zürich, 1953); J. L. Talmon, *The Origins of Totalitarian Democracy* (London, 1952); René Gillouin, *Man's Hangman Is Man* (Mundelein, Illinois, 1957); Walter Lippmann, *Essays in the Public Philosophy* (Boston, 1955); Lord Percy of Newcastle, *The Heresy of Democracy* (London, 1954); Hannah Arendt, "Authority in the Twentieth Century," *The Review of Politics* (October, 1956); P. Worsthorne, "Democracy v. Liberty," *Encounter* (January, 1956); Christopher Dawson, "The Birth of Democracy," *The Review of Politics* (January, 1957). Dawson quotes Thomas Paine as the American prophet of revolutionary democracy. In his famous *Common Sense* (1776), Paine exclaims: "We have it in our power to begin the world all over again. A situation similar to the present hath not happened since the days of Noah until now. The birth-day of a new world is at hand, and a race of men perhaps as numerous as all Europe contains are to receive their portion of freedom from the event of a few months." Fortunately, the influence of men like Alexander Hamilton, James Madison, and John Adams was strong enough to prevail against the Jacobin views of Thomas Paine and his disciples.

The phenomenon of "eternal Jacobinism," which, incidentally, is also an essential element of Communism, has not yet been sufficiently clarified. In particular, less attention than the subject merits has been paid to the characteristic idea of revolutionary new beginnings—as if all the preceding millennia had been waiting for our own higher illumination and purer intentions. Yet it should be obvious how important and fateful this idea is. There is an *a priori* presumption that it contains more than a little theology, and this is very plain in the works of Rousseau and also of Paine, as Dawson makes clear. A Catholic variant of this idea started with Lamennais, who imagines that in our day the darkness is suddenly being lifted from the peoples of the world and replaced by progress. Lamennais has become the pioneer of a Catholic progressivist movement, a Jacobinism of the Cross, which is softening up even the Catholic masses for totalitarian democracy and Communism—most of all in Lamennais' own country. In the orthodox Eastern churches the same heresy has found occasional supporters, for example N. Berdyaev.

19. On isolation of the individual in mass society, see A. Rüstow, "Vereinzelung," Vierkandt Festschrift *Gegenwartsprobleme der Soziologie* (Potsdam, 1949); Paul Halmos, *Solitude and Privacy* (London, 1952); and Riesman, *op. cit.*

20. The quotation is from Joachim Bodamer, *Gesundheit und technische Welt* (Stuttgart, 1955), 203. See also Hermann Friedmann, *Das Gemüt, Gedanken zu einer Thymologie* (Munich, 1956), and *Au service de la personne, médecine et monde nouveau* (Paris, 1959), a symposium of European physicians working under the leadership of Paul Tournier.

21. *Goethe's Faust,* an abridged version translated by Louis MacNeice (London, 1951), 283.

22. Charles Morgan, *Liberties of the Mind* (London, 1951), 111.

23. Apart from Bernanos, this subject has occupied many other contemporary writers, including T. S. Eliot and Evelyn Waugh, but I know of no penetrating analysis, except for an excellent chapter in Kirk's *A Program for Conservatives.* In a later work, *Beyond the*

271

Dreams of Avarice, Kirk repeatedly refers to the "Age of Boredom." The real discoverer of "social ennui" is probably Dean W. R. Inge (*A Pacifist in Trouble* [London, 1939]).

24. The quotation is from W. H. R. Rivers, *Essays on the Depopulation of Melanesia* (cited by Kirk in *A Program for Conservatives,* 104). One may wonder, incidentally, whether the detractors of romanticism are at all aware of the part played in the colored peoples' regrettable antagonism toward the West by their dislike of what they call Western "materialism." Their dislike of it is at least as great as their desire to emulate us and the envy that is caused by the difficulty of emulation.

25. Richard Kaufmann, *Süddeutsche Zeitung,* October 6-7, 1956, with reference to a survey on the "use of leisure in an industrial town" arranged by the Westphalian Institute of Journalism. The famous Kinsey Report, whose repulsive emphasis on the merely physical is in itself a sign of our times, also has its bearing on this subject, for what else is this rampant and obsessive sexuality but the expression of an infinitely bored society that is merely acerbating its boredom with such erotic degradation? More than anything else, this report demonstrates how our world is on the brink of dying of boredom like the Melanesians. With regard to Great Britain, Rowntree and Lavers, *op. cit.,* have painted a no less depressing picture of the boredom of mass society.

Women, to the extent that they remain interested in their households and children and perhaps also as a result of their different nature, have a better chance of compensation; there is a presumption, therefore, that modern mass society affects and bores men more than women (cf. Ludwig Paneth, *Rätsel Mann* [Zürich, 1946]; Bodamer, *op. cit.,* 49ff.). Finally, the question of how old people fare in this society has called forth a vast amount of literature. It is indeed a very important question and one that opens up grim perspectives. What happens to children has already been discussed in the text of this volume.

26. Colm Brogan, *The Democrat at the Supper Table* (London, 1945), 171.

27. Tocqueville, *op. cit.*, Vol. II, Bk. II, Ch. 15, p. 145. See also his observations in *ibid.*, Ch. 17.

28. Arnold Weber, "Zur Psychologie des Fernsehens," *Schweizer Monatshefte* (February, 1957). Weber is a professor of child psychiatry at the University of Bern.

29. In Switzerland, the revolt against technological utilitarianism has admirably expressed in Emil Egli's essay "Auftrag und Grenzen der Technisierung," *Die Schweiz*, Jahrbuch der Neuen Helvetischen Gesellschaft (1956).

30. Charles Baudouin (*op. cit.*, 21) raises the question of whether the vandalism of the last war's bombing is not appropriate to the spirit of irreverent and destructive modernism. In fact, there was no strategic (not to mention moral) justification for the destruction of ancient town centers. These town centers are a thorn in the flesh of *avant-garde* architects. A last and rather touching vestige of shame (or maybe a thought for tourist interest) has preserved the eighteenth-century British Governor's Palace in Boston amidst that city's concrete-and-glass canyons, although, of course, the building's nobility, set among oppressive dreariness and ugliness, is all the more eloquent proof of modern barbarity. It is significant that the stream of Americans which inundates Europe each year is not matched by an opposing stream of Europeans visiting America and that there is no European counterpart to the type of American weary of his continent and coming to settle in Europe, such as was depicted in some of Sinclair Lewis' novels. The ultimate reason is that Americans still find in the Old World some vitamin of the soul which they often miss in the New. It is in this light that the Europeans' sedulous imitation of America should be judged.

31. See some highly pertinent observations in Kirk, *Beyond the Dreams of Avarice*, 308-309. This is a further reason for the decline of the arts (see Note 16 above). I do not know whether any attempt

has been made to interpret modern surrealist and abstract art in this light, but it seems to me that the point of view stressed here should not be neglected if we are to understand a school of painting which, at best, produces something like wallpaper patterns. This point of view might help to develop in one important respect the excellent analyses by Hans Sedlmayr (*Die Revolution der modernen Kunst* [Hamburg, 1955]) and Wladimir Weidlé (*Les abeilles d'Aristée, Essai sur le destin actuel des lettres et des arts* [3rd ed., Paris, 1954]) If today's visual arts increasingly lose their "content" (see Max Picard, *Die Atomisierung der modernen Kunst* [Hamburg, 1954] and the impressive last chapter of Peter Meyer's *Europäische Kunstgeschichte* [Vol. II, Zürich, 1948]), the most obvious explanation is that they, too, are succumbing to the realities of modern mass society. We can summarize the situation by saying that the artist's position in our society cannot be assessed without being treated as part of our overall pathological picture. In this, the artist plays a dual role: (1) as a symptom of the decline of "bourgeois" society, that is, as an essential part of a process in which the artist appears as the prototype of rootlessness and, since the romantic age, even prides himself on his "creative" disdain of the despised Philistine; and (2) as a victim of the selfsame process, which increasingly turns creative art into business, which steadily diminishes the demand for genuine art—partly for technical reasons (photography, radio, films, television, better methods of reproduction, etc.) and partly for sociological ones (disappearance of patronage, mass culture, proletarianization, etc.)—and which subjects creative production to the modern laws of quantity and speed, and the artist to the law of supply and demand. We shall have more to say about this in another context.

32. *New York Times*, June 27, 1955. As regards Great Britain, a recent poll among 5,603 Cambridge undergraduates revealed that 11 percent of the men respondents and 34 per cent of the women had "decided" to emigrate after graduation and that 27 per cent of

the men and 15 per cent of the women had "considered" doing so. Even allowing for the fact that only 6 per cent of the total replied to these particular questions, the proportion is higher than that of emigrants in general to total population (*Economist* [February 9, 1957]). The reasons given—dissatisfaction, disgruntlement—correspond to those suggested in the text.

33. "The strength of the romantic current always corresponds precisely to the alienation of the exponents of a refined civilization from the general human base." (Meyer, *op. cit.*, 349)

CHAPTER III

1. Among contemporary economists who have turned their attention to the ethical framework of the economy, we may mention J. M. Clark, *The Ethical Basis of Economic Freedom* (The Kazanjian Foundation Lectures, 1955) and David McCord Wright, *Democracy and Progress* (New York, 1948). It is also pertinent to recall the following passages from J. C. L. Simonde de Sismondi's *Nouveaux principes d'économie politique* (2nd ed., Paris, 1827) : "The mass of the people, and the philosophers, too, seem to forget that the increase of wealth is not the purpose of political economy but the means at its disposal for insuring the happiness of all." (p. iv) "When England forgets people for thinking of things, is she not sacrificing the aim to the means?" (p. ix) "A nation where no one suffers want, but where no one has enough leisure or enough well-being to give full scope to his feelings and thought, is only half civilized, even if its lower classes have a fair chance of happiness." (p. 2) Indeed, the entire first chapter of this book is well worth rereading.

2. Cf. my two treatises, *Borgkauf im Lichte sozialethischer Kritik* (Köln and Berlin, 1954) and *Vorgegessen Brot* (Köln and Berlin, 1955).

3. Another apposite example of the progressive decline of the

significance of ownership and related norms and institutions is the deteriorating morale of debtors at the expense of creditors. This development can be observed in many countries. The courts are lenient in cases of default and bankruptcy, and the result is that the creditor is often deprived of his rights and property in the name of mistaken "social justice." It should hardly be necessary to point to the expropriation of landlords because of rent control in many countries or to the effects of progressive personal taxes.

4. From the characteristically plentiful recent literature, we may mention, apart from the works cited in Note 1 above: F. H. Knight, *The Ethics of Competition* (London, 1935); K. E. Boulding, *The Organizational Revolution* (New York, 1953); Daniel Villey, "The Market Economy and Roman Catholic Thought," *International Economic Papers* (No. 9, 1959); G. Del Vecchio, *Diritto ed Economia* (2nd ed., Rome, 1954); W. Weddigen, *Wirtschaftsethik* (Berlin-Munich, 1951); A. Dudley Ward (ed.), *The Goals of Economic Life* (New York, 1953); and D. L. Munby, *Christianity and Economic Problems* (London, 1956).

5. The relevant discussion has been fully reported by Carlo Antoni in A. Hunold (ed.), *Die freie Welt im kalten Krieg* (Erlenbach-Zürich, 1955).

6. The idea here expressed is treated more fully in my book *The Social Crisis of Our Time*, 225-27.

7. The economist who rejects utilitarianism finds himself in the distinguished company of J. M. Keynes, who has this to say about the Benthamite tradition: "But I do now regard that as the worm which has been gnawing at the insides of modern civilisation and is responsible for its present moral decay." (J. M. Keynes, *Two Memoirs* [London, 1949], 96) In connection with the passage from Macaulay's *Essays* mentioned in the text, we recall Bentham's remark: "While Xenophon was writing his history and Euclid teaching geometry, Socrates and Plato were talking nonsense under pretence of talking wisdom and morality." (Quoted from *Time and Tide* [May 19, 1956]) There is a clearly visible road from this kind

of Philistine utilitarianism to positivism and the philosophy of logical analysis.

8. For more detail, see my essay "Gegenhaltung und Gegengesinnung der freien Welt," in *Die freie Welt im kalten Krieg*, 183-211.

9. For an exposition of the overall problem, see my study "Unentwickelte Länder," *ORDO, Jahrbuch für die Ordnung von Wirtschaft und Gesellschaft* (1953), 63-113. For the topical case of the Arab world, see Walter L. LaQueur, *Communism and Nationalism in the Middle East* (London, 1957).

10. Apart from LaQueur's book, see also Emil Brunner, "Japan heute," *Schweizer Monatshefte* (March, 1955); Ramswarup, *Gandhism and Communism* (New Delhi, 1955), in which we find this statement: "Our intellectualized leftist conscience sees nothing but illiteracy, inadequacy, misery and frustration around and hopes to remove these by the blue-prints of 5-year plans. Gandhiji, on the other hand, brought a message of hope and suggested ways of improvement, not by destroying existing patterns but by bearing with them, by improving them." (p. 11); Harry D. Gideonse, "Colonial Experience and the Social Context of Economic Development Programs," in R. A. Solo (ed.), *Economics and the Public Interest* (New Brunswick, 1955); F. S. C. Northrop, *The Taming of Nations* (New York, 1952); Eugene Staley, *The Future of Underdeveloped Countries* (New York, 1954); M. R. Masani, "The Communist Party in India," *Pacific Affairs* (March, 1951).

11. Admirably apposite is Theodor Mommsen's summing up of the staleness of ancient Rome, which formed the background for a personage like Catiline: "When a man no longer enjoys his work but works merely in order to procure himself enjoyments as quickly as possible, then it is only an accident if he does not become a criminal." (Quoted from Otto Seel, *Cicero* [Stuttgart, 1953], 66)

12. Wilhelm Röpke, "A European Looks at American Intellectuals," *The National Review* (November 10, 1956). The literature on this important subject reflects the facts, for it is divided into the two extremes of anti-capitalist intellectuals on the one hand and

anti-intellectual capitalists on the other. This means that the problem as such is lost to view. This weakness is also apparent in F. A. Hayek (ed.), *Capitalism and the Historians* (Chicago, 1954), however valuable this book is in other respects as a corrective of our ideas about economic history. Cf. my review ("Der 'Kapitalismus' und die Wirtschaftshistoriker") in *Neue Zürcher Zeitung*, No. 614 (March 16, 1954).

13. On the ethical "middle level" of the market economy, see M. Pantaleoni, *Du caractère logique des différences d'opinions qui séparent les économistes* (Geneva, 1897); Wilhelm Röpke, *Die Lehre von der Wirtschaft* (8th ed., Erlenbach-Zürich, 1958), 41-46; and *idem, Internationale Ordnung—heute*, 116-35.

14. Cf. my essay "Unentwickelte Länder," *ORDO, Jahrbuch für die Ordnung von Wirtschaft und Gesellschaft* (1953).

15. How Christianity overcame this phase in the course of its development as a dogma and a church and how it came once more to acknowledge the cultural value of "loving oneself" is very evident from Augustine's example. Cf. Hans von Soden, *Urchristentum und Geschichte* (Tübingen, 1951), 56-89.

16. Lord Acton, *The History of Freedom and Other Essays* (London, 1907), 28.

Lord Acton was a Catholic and might well have invoked St. Thomas Aquinas: "Ordinatius res humanæ tractantur, si singulis immineat propria cura alicuius rei procurandæ; esset autem confusio, si quilibet indistincte quælibet procuraret." (*Summa Theologiae*, II, II, 66, 2. Quoted from Joseph Höffner, "Die Funktionen des Privateigentums in der freien Welt," in E. von Beckerath, F. W. Meyer, and A. Müller-Armack (eds.), *Wirtschaftsfragen der freien Welt* [Erhard-Festschrift, Frankfurt a. M., 1957], 122.)

We might also recall the Pilgrim Fathers, the first English colonizers of New England, who, devout Calvinists as they were, thought they could set up a purely communist system of agriculture; but a few years later, they were forced by the catastrophic decline in yields to change over to a market system and private ownership.

17. The part played by the art collector's passion in the lives of American multimillionaires of the past generation is described in an entertaining biography of the art dealer who supplied them: S. N. Behrman, *Duveen* (London, 1952). Their names are immortalized in the art galleries they created, which include the National Gallery in Washington, the Frick Gallery, and special collections at the Metropolitan Museum in New York. It seems as if this back door to immortality was one of Duveen's most effective selling points.

On the other hand, even supreme intellectual achievements are not always free from the profit motive, as Goethe's example shows. It seems that it was an attractive offer by his publisher, Cotta, which finally led Goethe to complete his *Faust*. Schiller had solicited this offer behind Goethe's back; we have his letter to Cotta of March 24, 1800: "I am afraid Goethe will completely neglect his Faust, into which so much work has already gone, unless some stimulus from outside in the form of an attractive offer stirs him to take up this great work once more and finish it. . . . However, he expects a large profit, for he knows that this work is awaited with suspense in Germany. I am convinced that you can get him, by means of a brilliant offer, to complete this work in the coming summer." Goethe's prompt reaction can be seen in his letter to Schiller of April 11, 1800. But who would therefore deprecate the profit motive?

18. "The Benthamite delusion that politics and economics could be managed on considerations purely material has exposed us to a desolate individualism in which every man and every class looks upon all other men and classes as dangerous competitors, when in reality no man and no class can continue long in safety and prosperity without the bond of sympathy and the reign of justice." (Russell Kirk, "Social Justice and Mass Culture," *The Review of Politics* [October, 1954], 447) If we want to understand fully this error of liberal immanentism, which we first meet in such disarming purity in Say's youthful work *Olbie*, then in the writings of Bentham and his school, and which had a last bright flicker in Herbert Spencer's work, we must remember that at that time the liberation from really

constrictive bonds was an absorbing task, while the moral reserves were still intact enough to be taken for granted. A similar situation existed in Germany after 1945, when it was necessary to give priority to the need of overcoming intolerable poverty by releasing the economic forces weakened by repressed inflation. The one-sidedness of nineteenth-century individualism was paralleled by the equally conspicuous one-sidedness of political individualism, whose fatal ideal of unitarian democracy can be understood as a reaction to the pluralistic petrification of the *ancien régime*.

The roots of the moral blindness of individualism and utilitarianism naturally reach far back into the eighteenth century, to Helvétius, Holbach, Lamettrie, and D'Alembert, just as its ramifications ultimately reach forward to Marx and Engels.

19. Cf. my book *International Economic Disintegration* (3rd ed., London, 1950), 67ff., and my course of lectures on "Economic Order and International Law" at The Hague Academy of International Law (*Recueil des Cours 1954* [Leiden, 1955]).

20. Cf. Goetz Briefs, "Grenzmoral in der pluralistischen Gesellschaft," in *Wirtschaftsfragen der freien Welt*.

21. The problems of competition and the dilemma it so often involves can be studied very well in the example of universities. If one knows the system of those countries where the lecturer draws attendance fees and therefore has a financial interest in the outward success of his lectures, one realizes how poisonous an atmosphere of rivalry can thus be created and how the teacher is tempted to court outward success more than is right and proper. On the other hand, this system provides a good stimulant for weaker characters who are not sufficiently conscious of the obligations of their office.

22. The idea of *nobilitas naturalis* is, of course, so old that it is difficult to trace its spiritual genealogy. It may be worth noting, though, that the idea was quite familiar to a democrat like Thomas Jefferson, who is above any suspicion of reactionary opinions. On October 28, 1813, Jefferson wrote to John Adams, who was a conservative: "I agree with you that there is a natural aristocracy among

men. The grounds of this are virtue and talents. . . . The natural aristocracy I consider as the most precious gift of nature, for the instruction, the trusts and government of society. And indeed it would have been inconsistent in creation to have formed man for the social state, and not to have provided virtue and wisdom enough to manage the concerns of society." (A. Koch and W. Peden, *The Life and Selected Writings of Thomas Jefferson* [Modern Library, New York], 632-33) The application to the particular case of the market economy can be found in my book *The Social Crisis of Our Time*, 134ff. See also Wright, *op. cit.*, 25ff.

23. We again quote an author beyond suspicion: "[The legislator] has not fulfilled his task if, in his desire to insure equal satisfaction of all needs, he renders impossible the full development of outstanding individuals, if he prevents anyone from rising above his fellows, if he cannot produce anyone as an example to the human race, as a leader in discoveries which will benefit all." (Simonde de Sismondi, *op. cit.*, II, 2) The same idea is forcefully expressed by Alexis de Tocqueville in his *Democracy in America*. See also L. Baudin, "Die Theorie der Eliten," in *Masse und Demokratie*, 39-54.

24. H. K. Röthel, *Die Hansestädte* (Munich, 1955), 91.

25. W. H. Hutt, *Economists and the Public* (London, 1936); Wilhelm Röpke, "Der wissenschaftliche Ort der Nationalökonomie," *Studium Generale* (July, 1953).

26. See my *Mass und Mitte*, 200-218. Since that book was published, I have become even more firmly convinced that advertising, in all of its forms and with all of its effects, one of the foremost of which is to encourage the concentration of firms, is one of the most serious problems of our time and should receive the most critical attention of those few who can still afford to speak up without fear of being crushed by the powerful interests dominating this field. However, the interested parties are likely to put up fierce resistance, as we know from experience. To give a sample of it, and at the same time to illustrate the point of view developed in the text, I quote the following sentences from an article against the limitation of outdoor

advertising: "With all due respect to the tidiness of our towns and landscapes and to the need of protecting monuments of nature, art, and culture, the aesthete's susceptibilities must today yield to the very concrete claims of life. . . . Undoubtedly, town and country would be prettier without posters and obviously also without the rush of traffic and all the other well-known inevitable troubles and distinguishing marks of modern business activity. But all of this, whether good or unpleasant, cannot be painted out of modern public and business life by, as it were, faking the picture with the brush of a buildings-preservation policy." (*Niedersächsische Wirtschaft* [July 20, 1954]) It is hardly possible to state more crudely an opinion whose power is, unfortunately, only too easy to imagine.

27. See Note 2 above and the works to which it refers.

28. On East-West trade, see Wilhelm Röpke, "Aussenhandel im Dienst der Politik," *ORDO, Jahrbuch für die Ordnung von Wirtschaft und Gesellschaft* (1956), 45-65.

29. On economic policy in a mass democracy, see Lippmann, *op. cit.*; Felix Somary, *Democracy at Bay: A Diagnosis and a Prognosis* (New York, 1952); Lord Percy of Newcastle, *op. cit.*; Gillouin, *op. cit.*; Kirk, "Social Justice and Mass Culture," *The Review of Politics* (October, 1954); and Wright, *op. cit.* Modern "television democracy" is the nadir of the downward development so far.

30. Bertrand de Jouvenel, *Du Pouvoir, histoire naturelle de sa croissance* (Geneva, 1945), 390ff.

31. Benjamin Constant, *Oeuvres politiques* (ed. Louandre, Paris, 1874), 248f.

32. On pressure groups, see Boulding, *op. cit.*, and A. Rüstow, *Ortsbestimmung der Gegenwart* (Erlenbach-Zürich, 1957), Vol. III, 171ff. On the "labour standard," see J. R. Hicks, "Economic Foundations of Wage Policy," *Economic Journal* (September, 1955), 391.

33. Wilhelm Röpke, *Wohnungszwangswirtschaft—ein europäisches Problem* (Düsseldorf, 1951); M. Friedmann and George J. Stigler, "Roofs or Ceilings?" *Popular Essays on Current Problems*, Vol. I, No. 2 (September, 1946); Alfred Amonn, "Normalisierung

der Wohnungswirtschaft in grundsätzlicher Sicht," *Schweizer Monatshefte* (June, 1953). A comprehensive postwar exposition of this hair-raising chapter of economic policy has yet to be written. The true state of affairs became clear to me recently when I received a letter from a German socialist politician. He wrote that by now everybody was of one mind about this troublesome matter but that he would be glad to hear from me concerning what one could do about it in practice. I replied that that was not my business but his. I expected of him, I said, that he should openly defend in public the view which he had expressed in his letter to me, and I proposed that as a beginning we publish our correspondence. I received no reply.

34. The prototype of the modern economocrat is the eighteenth-century physiocrat. The physiocrats—or *économistes,* led by Quesnay—are clearly the ancestors of all the power-thirsty, cocksure, and arrogant planners and organizers. Walter Bagehot (*Biographical Studies* [London, 1881], 269f.) paints a vivid picture of them. He says that a contemporary of Quesnay's wrote of him that he was convinced that he had reduced economic theory to a mere calculation and to axioms of irrefutable evidence. Tocqueville (*L'ancien régime et la révolution* [1856], Chapter 3) says of the physiocrats: "They not only abhor certain privileges, but all diversity: they would worship equality even if it meant general slavery. Whatever does not fit in with their designs has to be smashed. They have little respect for contracts and none for private rights; or rather, they do not, strictly speaking, admit private rights at all, but only the common benefit."

CHAPTER IV

1. An extensive discussion of the Beveridge Plan is to be found in my book *Civitas Humana* (London, 1948), 142-48. In his *Full Employment in a Free Society* (London, 1944), the creator of this famous plan, by means of which Great Britain became the model of

the welfare state, subsequently contributed much to complementing the egalitarian ideology of the welfare state with the ideology of inflationary "full employment." See the pertinent critique of this second Beveridge plan in Henry C. Simons, "The Beveridge Program: An Unsympathetic Interpretation," *Journal of Political Economy* (September, 1945), reprinted in *Economic Policy for a Free Society*, 277-312; Lionel Robbins, *The Economist in the Twentieth Century* (London, 1954), 18-40. Both critics come to the correct conclusion, now confirmed by facts, that the full-employment policy advocated by Beveridge must result in inflation. It is much to Lord Beveridge's credit that he himself later repeatedly and frankly criticized the development which his first plan set in motion. In his later book *Voluntary Action* (London, 1948), for instance, he took occasion to place voluntary group aid in its proper light. However, he seems never to have realized how great a part he played in the development he criticized. Not long ago, he frankly declared in a lecture that inflation was destroying the savings which he had set aside for his own old age; it may therefore happen, he said, that he would live longer than he could afford to. But he does not seem to have grasped that a large part of the responsibility for this inflation, which erodes his savings and threatens his carefree remaining years, belongs to his own creation, the welfare state, together with overfull employment, also a subject of his praise. He appears as the pathetic figure of a man who does not know that he himself sawed off the branch on which he sat.

2. Colin Clark, *Welfare and Taxation* (Oxford, 1954); A. C. Pigou, "Some Aspects of the Welfare State," *Diogenes* (July, 1954); Bertrand de Jouvenel, *The Ethics of Redistribution* (Cambridge, 1951); Hans Willgerodt, "Die Krisis der sozialen Sicherheit und das Lohnproblem," *ORDO, Jahrbuch für die Ordnung von Wirtschaft und Gesellschaft* (1955), 145-87.

3. Lionel Robbins, "Freedom and Order," in *Economics and Public Policy* (Washington, 1955), 152. A few lines earlier, Robbins says: "In a society in which incentive and allocation depend on

private enterprise and the market, a continuous redistribution of income and property in the interests of a pattern of equality, or something approximating to equality, is almost a contradiction in terms."

4. Helmut Schoeck, "Das Problem des Neides in der Massendemokratie," in *Masse und Demokratie*, 239-72.

5. "'The hatred that men bear to privilege increases in proportion as privileges become fewer and less considerable, so that democratic passions would seem to burn most fiercely just when they have least fuel. I have already given the reason for this phenomenon. When all conditions are unequal, no inequality is so great as to offend the eye, whereas the slightest dissimilarity is odious in the midst of general uniformity; the more complete this uniformity is, the more insupportable the sight of such a difference becomes. *Hence it is natural that the love of equality should constantly increase together with equality itself, and that it should grow by what it feeds on.* This never dying, ever kindling hatred which sets a democratic people against the smallest privileges is peculiarly favorable to the gradual concentration of all political rights in the hands of the representative of the state alone. The sovereign, being necessarily and incontestably above all citizens, does not excite their envy, and each of them thinks that he strips his equals of the prerogatives that he concedes to the crown." (Alexis de Tocqueville, *Democracy in America*, Vol. II, Book IV, Chapter 3, p. 295. My italics.)

6. A discussion of the demand for equality and all of its consequences is to be found in my earlier book *Mass und Mitte*, 65-75. I still subscribe to that critique, as well as to my serious misgivings about that subtle and therefore most tempting form of equality which goes by the name of equality of opportunity. The arguments I put forward then would seem to be convincing enough, in particular the argument that it would be completely arbitrary to aim at equality of opportunity only in material matters susceptible to the leveling action of the state, while inequality must be accepted in other fields —unequal health, unequal intelligence, unequal character. If, there-

fore, opportunities are to be really the same, the material conditions (the income and wealth of the child's parents) must be measured out in such doses that they add up with the non-material and non-equalizable conditions to "equal opportunity." Suppose a child has poor health but his parents can, at least, equip him with better material conditions for the struggle of life. Now what possible justification is there in depriving him even of these? Should not the others be glad to have inherited a healthy stomach, a sound heart, or nerves of steel? How is all that to be calculated? In this light the incessant redistribution which strict equality of opportunity presupposes looks even more outrageous than it does in any case. Furthermore, if it is just that a man may own private property—and the advocates of equality of opportunity fortunately do not go so far as to deny it— why should it be unjust that his children benefit by it? I may do anything I like with my income and wealth—I may build a house, buy a television set, acquire a luxury car, travel around the world— only one thing I may not do, namely, give my children the best and most careful education. For the rest, we shall see in the next chapter that the claim for equality of opportunity corresponds to an extreme ideal of liberalism according to which a continuous race of all for everything is desirable. The question arises: by what right is this race to stop at the national frontiers?

7. Heinrich Heine, *Deutschland,* Kaput I. It should be obvious that the collective utilitarianism and epicureanism of the welfare-state ideology is closely connected with the disappearance of belief in transcendence and immortality. Cf. Aloys Wenzl, *Unsterblichkeit* (Bern, 1951).

8. Colm Brogan, *The Educational Revolution* (London, 1954), paints a vivid picture of Great Britain, which, for the time being, remains an extreme case. In the United States, "classless," as applied to education, is a make-believe, as in many other fields, since parents are free to send their children to private schools if they want them to have a better education than can be expected in the public schools. The only drawback is that this is far more expensive than

the much-maligned school fees current in European countries for good public schools. We possess a vast documentation concerning the appalling deterioration of the educational level entailed by socialization of education. Another important factor is that if so many young people go to the universities, the non-academic groups of the population are continuously deprived of their most intelligent and enterprising elements (Erik R. von Kuehnelt-Leddihn, *Freiheit oder Gleichheit?* [Salzburg, 1953], 473) and family ties are disrupted. See also Note 1 to Chapter V of this volume.

9. Hermann Levy, *National Health Insurance: A Critical Study* (London, 1944); M. Palyi, *Compulsory Medical Care and the Welfare State* (Chicago, 1950); F. Roberts, *The Cost of Health* (London, 1952); Werner Bosch, *Patient, Arzt, Kasse* (Heidelberg, 1954); H. Birkhäuser, "Der Arzt und der soziale Gedanke in der Medizin," *Schweizerische Medizinische Wochenschrift*, No. 5 (1956).

10. Concerning the National Health Service, a distinguished British economist writes: "The important economic question about that scheme was this: if there is a service the demand for which at zero price is almost infinitely great, if no steps are taken to increase the supply, if the cost curve is rising rapidly, if every citizen is guaranteed by law the best possible medical service and if there is no obvious method of rationing, what will happen? I do not recall any British economist, before the event, asking these simple questions." (J. Jewkes, in *Economics and Public Policy*, 96) Compare this with the observation by M. Palyi (*op. cit.*, 71): "The abolishment altogether of a compulsory sickness scheme, once established, even if bankrupt and unsatisfactory, is beyond imagination. It has never happened." A further testimony: "Enthusiasts for nationalized medicine found themselves in competition with the enthusiasts for extended education, state subsidized housing, higher state pensions and benefits, and a dozen other schemes with a strong emotional and vote-catching appeal. . . . I believe that the contemporary and scientific conception of medicine cannot flourish fully and firmly where

medicine has been socialized." (Colm Brogan, "The Price of Free Medicine," *The Freeman* [June, 1956]) Finally, a British physician confirms this view: "The cost to the country in money is easily expressed, understood, accepted, amended or rejected. The cost to the country in health and happiness which will result from the degradation of doctors is beyond our powers of comprehension." (Scott Edward, "Retreat from Responsibility," *Time and Tide* [October 10, 1953])

11. On the illusion of the welfare state, see Colin Clark, *op. cit.*, and M. J. Bonn, "Paradoxien eines Wohlfahrtsstaates," *Aussenpolitik* (April, 1953).

12. This is, among other things, what Colin Clark's proposals come to. Compare the following recent report from Belgium (*Neue Zürcher Zeitung*, No. 1209 [April 27, 1957]). The socialist Minister of Labor proposed, by means of the method now fashionable everywhere, to raise the income limit of compulsory state insurance and to merge the various private pension funds into a state fund. The result was a storm of indignation among the workers and trade-unions. The social charges of Belgian industry had risen in twelve years from 25 per cent of the wage bill to 41 per cent, and the Belgian workers and employees decided that this was enough—more than enough. They asked such awkward questions as whether there was still any reasonable relationship between the growing social contributions and actual services and whether there were no cheaper ways of obtaining old-age insurance.

13. It is normal to deplore the fate of men like Winckelmann, Herder, Hebbel, Racine, and many others whose genius was handicapped by poverty, but the point is that all of them succeeded in coming to the top, thanks to the diversified structure of society in their time. Encouragement and help were to be had in many places and from many people: the master of a school, a princely patron, a secretarial post, a hospitable country mansion. In these circumstances there was a very high probability of being able to set one's foot on the rung of some ladder; at any rate, to say the least, this

probability could well stand comparison with the likelihood that no genius will go unnoticed in the present welfare state. How the rise of talent was possible at that time in the most adverse circumstances is impressively seen in Winckelmann's life. (C. Justi, *Winckelmann und seine Zeitgenossen* [2nd ed., Leipzig, 1898], I, 22 and 28)

Many other biographies testify to the same thing. Take, for instance, the life of Scharnhorst, a tenant farmer's son from Hannover, who received tuition in mathematics from a retired major (this happened in my own village) and was then sent to a small military college by the Count of Schaumburg-Lippe. One cannot help being both touched and astonished by the climbing feats of these men as they rose, from one foothold to the next, in society. It is not so certain that the socialized chairlift of the welfare state always achieves the same successes. In other respects, too, our age of the welfare state has little reason to consider itself so superior to the social hardships of the past. A little more modesty is indicated in relation to our forebears. Anyone who, like myself, has grown up in the simple conditions of a village can easily remember the time when the different classes stood together in a neighborly way, whereas today they are far removed from each other. The real inequality of men has not diminished but has increased during the last one hundred years. As an example, take Zelter, who started out as a builder's apprentice and ended up as a professor of music and close friend of Goethe without losing contact with his own milieu. "A life of this kind," Paulsen wrote as long ago as the end of the last century (*Ein System der Ethik* [2nd ed., Berlin, 1891], 727), "would be inconceivable today. Nowadays Zelter would have gone through secondary school and studied architecture, he would have learned to draw and calculate, would have taken mechanics and history of art, and he would have become an architect and officer of the reserve and would never have built a single wall. He would have been an employer of masons, not their fellow and instructor. Or else he would have remained a mason and a fellow of masons, but then he

289

would not have gained the friendship of a *Geheimrat* and minister, nor become a professor of music."

14. I entirely agree with the incisive criticism recently expressed by two distinguished contemporary economists: F. A. Hayek, "Progressive Taxation Reconsidered," in Mary Sennholz (ed.), *Freedom and Free Enterprise: Essays in Honor of Ludwig von Mises* (New York, 1956), 265-84; Wright, *op. cit.*, 94ff.

15. Cf. Chester C. Nash, "The Contribution of Life Insurance to Social Security in the United States," *International Labour Review* (July, 1955). For corresponding Swiss figures, see E. Marchand, "Le développement de l'assurance en Suisse," *Journal des Associations Patronales 1906-1956*. In 1953, the last year for which figures are available, the sums paid out by insurance companies exceeded the payments of old-age and widows' insurance by nearly one hundred million francs.

16. The idea that economic laws exclude the possibility of the broad masses' providing for themselves by means of the accumulation of property and that these same economic laws make this self-provision a privilege of the few could grow only in the soil of popular Keynesianism. I discussed this in more detail in my articles "Probleme der kollektiven Altersversicherung" (*Frankfurter Allgemeine Zeitung*, February 25, 1956) and "Das Problem der Lebensvorsorge in der Freien Gesellschaft" (*Individual- und Sozialversicherung als Mittel der Vorsorge* [Bielefeld, 1956]), and I am pleased to quote now the clear and careful analysis by Hans Willgerodt, "Das Sparen auf der Anklagebank der Sozialreformer," *ORDO, Jahrbuch für die Ordnung von Wirtschaft und Gesellschaft* (1957), 175-98. Note also his comments on the pay-as-you-go system, with which I am in full agreement.

17. The problem of a sliding scale of pensions—called "dynamic pensions" in Germany and introduced, in somewhat modified form, in the spring of 1957—is discussed in my essays referred to in the preceding note. See also H.-J. Rüstow, *Zur volkswirtschaftlichen Problematik der dynamischen Sozialrente* (Berlin-Munich, 1956);

Das Problem der Rentenreform, Aktionsgemeinschaft soziale Markt-wirtschaft, Tagungsprotokoll No. 6 (Ludwigsburg, 1956).

18. Morgan, *op. cit.*, 122.

19. This particular form of the claim for international justice is all but forgotten today. I discussed it with the severity it deserves in my book *Internationale Ordnung—heute*, 164ff.

20. See my *L'économie mondiale aux XIXᵉ et XXᵉ siècles*, 165-220.

21. A typical work is G. Myrdal, *An International Economy* (New York, 1956). (See also P. T. Bauer's review of Myrdal's book, *Economic Journal* [March, 1959], and my own critical comments in *Wirtschaftsfragen der freien Welt*, 493ff.) The part played by plain envy in this, as in the national welfare state, is rightly stressed by Helmut Schoeck, "Der Masochismus des Abendlandes," in A. Hunold (ed.), *Europa—Besinnung und Hoffnung* (Erlenbach-Zürich, 1957).

22. Röpke, *Internationale Ordnung—heute*, 118, 133, and 241.

23. We get a good idea of the new interpretation of modern economic historians in F. A. Hayek (ed.), *Capitalism and the Historians*. However, as I have shown in my review of this book ("Der 'Kapitalismus' und die Wirtschaftshistoriker," *Neue Zürcher Zeitung*, No. 614 [March 16, 1954]), the pendulum now swings too far the other way. We may work out that the proletarians of that time ate more meat and drank more beer than we had so far thought and that materially things were only half as bad (though in my view even half would have been bad enough), but the crucial fact remains that they were proletarians in the widest and most unpleasant meaning of the word and that it was the first time in history that masses of them took the stage, together with their counterpart, the "capitalists." Modern economic and social historians would be well advised to share and analyze the contemporary witnesses' rather creditable feeling that this was a catastrophe. A large and by no means unimportant part of our own cultural crisis goes back to that time, and we cannot simply turn black into white and a minus into a plus.

This cannot be passed over in silence without leaving the most important aspect of the discussion unilluminated.

24. Wilhelm Röpke, "Offene und zurückgestaute Inflation," *Kyklos* (1947), 1; *idem.*, "Repressed Inflation," *Kyklos* (1947), 3. One of the harshest features of repressed inflation are the monetary "reforms" by which accumulated excess purchasing power due to the "repression" of inflation is periodically removed by calling in and converting cash holdings. Repressed inflation is nowadays really successful and comprehensive only in Communist countries because only they are ruthless enough and possess the required omnipotence of state and police. The example of Poland shows that any curtailment of this omnipotence breaks the dam. We are reminded that the National Socialist system of repressed inflation collapsed in 1945 at the same time as the political regime. Inflation, in its worst form of repressed inflation, is endemic in all Communist countries and indeed follows necessarily from the nature of the economic system, which is not so in the Western world. Inflation and collectivism are inseparable. (Cf. Wilhelm Röpke, *The Problem of Economic Order* [Cairo, 1951], 29-35)

25. A. J. Brown, *The Great Inflation 1939-1951* (London, 1955).

26. Irving Babbitt, *Democracy and Leadership* (Boston, 1924), 205-209.

27. Wilhelm Röpke, "Alte und neue Ökonomie," in *Wirtschaft ohne Wunder*, 66-96.

28. On the gold standard, see my *Internationale Ordnung— heute*, 110ff.

29. For a more detailed analysis, see my articles "Das Dilemma der importierten Inflation," *Neue Zürcher Zeitung*, No. 2128 (July 28, 1956), and "Nochmals: das Dilemma der importierten Inflation," *Neue Zürcher Zeitung*, No. 2798 (October 7, 1956). What I said there still seems valid. Imported inflation raises fewer problems inasmuch as it is a very obvious and accessible source of inflation. On the other hand, recent experience has shown it to be exceedingly difficult to deal with because there are great obstacles to the two

methods by which it could be removed, that is, devaluation in the country with stronger inflationary pressure and revaluation in the country with weaker inflationary pressure. Devaluation is resisted by governments because it is a reflection on their political prestige; revaluation is resisted by all of the economic interests which stand to suffer thereby.

30. The attentive observer of contemporary politics will find plenty of occasions to note the forces at work. The most recent example, and the one most familiar to German-speaking readers, is the German pensions reform (see Note 17 above) and the manner in which it was steam-rollered through Parliament. It was a most depressing spectacle to see how lightly government and Parliament took the immense responsibility which this measure implied. The law overrides all of the urgent recommendations to reverse course and to stimulate self-provision and individual responsibility. On the contrary, it is likely to impair saving considerably, not only because compulsory public provision is expanded, but also because the latter is largely put on a pay-as-you-go basis. Those who are responsible for this blow to saving have neither the excuse of not knowing the contrary arguments in good time nor of having seriously disproved them. No account at all was taken of the important and justified question of whether pensioners should not rather benefit from the economy's genuine productivity increase by means of an increase in their real income, that is, in the form of price reductions.

31. It would be rewarding to examine this in the greatest possible variety of fields, such as for example, in the mounting cost of the upkeep of historic monuments. It has become a serious problem to keep a city like Venice from falling into decay, although modern mass tourism does help. Nearly all of the articles of the antique trade or old Oriental carpets, etc., are steeply rising in price, and although this is in part due to chronic inflation, the secular rise in the prices of handmade goods plays its part, too. A similar argument applies to the real estate market. Here again the high prices

are in part due to chronic inflation, but in part they reflect the natural price increase of a good made scarce by population growth and urbanization in industrial countries, especially such circumscribed ones as Switzerland and Germany.

32. One of the earliest and clearest expositions of these relationships is to be found in Joseph A. Schumpeter, "The March into Socialism," *American Economic Review* (May, 1950). Note that Lord Beveridge *(Full Employment in a Free Society)* recommended overfull employment as an ideal state of affairs. "He demands a sellers' market for labor—a continuous excess of vacant jobs over idle hands—which obviously invites flight into assets via the labor market. Since it would mean an inflationary spiral of wage-rate increase even in the absence of any labor organization, Beveridge is naturally solicitous lest the trade-unions make demands which would frustrate efforts to sustain the value of money. . . . To expect labor monopolies not to demand monopolistic wages is, under any circumstances, unrealistic. To ask, with Sir William, that they use their power to keep wage rates below the competitive level is quixotic." (Simons, *op. cit.*, 291f.)

33. B. C. Roberts, "Towards a Rational Wages Structure," *Lloyds Bank Review* (April, 1957), 5.

34. To show how many errors are still current, even in circles where one would expect a more informed judgment, I quote Jeanne Hersch, an intelligent and lovable social philosopher who, in her *Idéologies et réalités* (Paris, 1956), 40, accuses me of some sort of moral defect because I think that a minimum of unemployment is necessary. This is precisely the kind of economically ignorant moralism which I had in mind and commented on earlier.

35. Central-bank policy might conceivably be complemented, or even replaced, by the immobilization of budget surpluses as a means to remove excess demand. But to discuss this question would take us too far here. It is connected with the yet more general question of whether public finance should take over the function of monetary control, which has hitherto been fulfilled by the central bank's credit

policy. This could be done by means of appropriate changes in taxation and expenditure, deflation to be met by deficit spending and inflation by the accumulation of unspent budget surpluses. I am ashamed to say that I must take my share of the blame for creating this concept of "functional finance" (*Krise und Konjunktur* [1932] and my subsequent book *Crises and Cycles* [1936]), but I am forced to admit now that it has stood the test neither of counter-arguments nor of experience. I agree entirely with the devastating criticism by Melchior Palyi (*Commercial and Financial Chronicle* [April 18, 1957]) and Friedrich A. Lutz (*Notenbank und Fiskalpolitik*, a lecture published by the Landeszentralbank von Baden-Württemberg in 1957). That part of the concept which is topical today, namely, the accumulation of budget surpluses as a means of combating inflation, is invalidated, at the very least, by the unrealistic assumption that in a modern democracy any budget surplus could be protected against the parliamentary appetite for larger expenditure.

36. Jacques Chastenet, *L'enfance de la Troisième 1870-1879* (Paris, 1952), 29 and 80, where the original sources are quoted.

37. "The *franc Germinal*, whose weight and fineness were not subject to government interference, was not only the instrument of French prosperity, but one of the solid bases of France's prestige. It is neither Bergson's putting intuition above reason nor the European nations' fratricidal wars which caused the real downfall of the edifice built, largely by the French, during the eighteenth and nineteenth centuries. This edifice collapsed when the inviolability of money ceased to be an article of faith, when governments assumed power over money and imagined they had the divine faculty of making something out of nothing." (Jacques Chastenet, *La France de M. Fallières* [Paris, 1949], 127)

38. I know of no comprehensive attempt to think through all of the implications of "simmering permanent inflation." Some salient points are discussed in F. A. Lutz, "Inflationsgefahr und Konjunkturpolitik," *Schweizerische Zeitschrift für Volkswirtschaft und Statistik* (June, 1957).

CHAPTER V

1. Here is one of these Jacobin opinions: "We proscribe the regional spirit, whether of the department or the commune; we hold that it is odious and contrary to all principles that some municipalities should be rich and others poor, that some should have vast possessions and others nothing but debts." (*Mémoires de Carnot*, I, 278, quoted from H. Taine, *La révolution*, III, 107) "We want no more local interests, memories, dialects, or patriotism. There must be only one bond among individuals, namely that which ties them to society as a whole. We shall break all the others; we cannot tolerate individual groupings, and we shall do our best to disintegrate the most tenacious of them all, the family." This is how Taine acutely sums up this Jacobin ideology. It is no accident that Carnot is the man who later became the creator of the mass armies based on compulsory military service. Few other institutions are so conducive to centralization and concentration of power; Bertrand de Jouvenel (*Du Pouvoir, histoire naturelle de sa croissance*, 11ff.) says it results in a modern Minotaur. A democracy inspired by the Jacobin myth of the sovereignty of the people rather than by the liberal idea that those governed should control government is bound to develop into a centralist "democratic despotism." There is fairly general agreement on this point today, but a little more alertness is called for to detect the underlying social philosophy in the contemptuous talk of the detractors of federalism, small nations, or small firms. We should look upon this kind of talk, which is now fashionable among so-called progressives, as a half-open door through which we get a glimpse of a house furnished in the Jacobin-Napoleonic style.

2. The term "liberalism" may be interpreted in a number of ways. In Switzerland, for example, political parties call themselves liberal when they are just as much conservative, in the sense in which Jacob Burckhardt and Alexandre Vinet may be called both conservative and liberal. Liberalism is the basic concept of the Swiss state, and anyone who defends it today against collectivist tendencies calls

himself a liberal. In Italy, the liberals are, on the one hand, anti-collectivist conservatives and, on the other, anti-clerical progressives who are anxious not to lose their connections with the Left. In Germany, the government's policy is liberal, but its chief exponent is a party which calls itself Christian Democratic, while the "liberals" are in the opposition. Such concepts are comparable to a musical instrument of a certain compass: in the highest and lowest registers it reaches into the range of another instrument, as does the viola into the ranges of the violin and the violoncello, but we still associate with each instrument the idea of a definite range which characterizes its sound. Thus the concept of liberalism has, in Europe, a considerable breadth within which its significance fluctuates. Much the same is true of America, except that the compass there is shifted considerably to the Left. Certain border notes are common to both variants of liberalism, but the average range is so different in America and Europe that the two concepts are almost the opposite of each other. The American associates with liberalism mostly notes which we in Europe would associate with the Social Democratic register. The New Deal, trade-unionism, planning, centralism, inflationary policies, radical taxation of income and wealth—all that is known as "liberal" in America, though this term certainly covers a lot of things which we in Europe would call by the same name. Confusion is even worse confounded by the fact that the concept is usurped by people and movements that are distinguished from Communists only by pretending that they are not.

3. No less a man than Proudhon has said the same thing: "Thus the systems of centralization, imperialism, communism, absolutism—all of these words are synonymous—derive from popular ideals. In the social contract, as conceived by Rousseau and the Jacobins, the citizen divests himself of his sovereignty; the town council, the departmental and provincial administrations are absorbed by central authorities and are no more than agencies under the direct control of the ministry. . . . State power invades every sphere, lays its hands on everything and usurps everything finally,

forever: army and navy, administration, jurisdiction, police, education, public building; banks, stock exchanges, credit, insurance, public assistance, saving, charity; forests, canals, rivers; religion, finance, customs, trade, agriculture, industry, transport. *And on top of everything heavy taxation, which takes one-fourth of the nation's gross social product.*" (*Du principe fédératif* [Paris, 1863], 69) I have italicized the last sentence in order to direct the reader's attention to Proudhon's perspicacity. It is easy to see why the centrist Marx hated this decentrist from the bottom of his heart.

4. "In America I saw the freest and most enlightened men placed in the happiest circumstances that the world affords; it seemed to me as if a cloud habitually hung upon their brow, and I thought them serious and almost sad, even in their pleasures. . . . Their taste for physical gratifications must be regarded as the original source of that secret disquietude which the actions of the Americans betray and of that inconstancy of which they daily afford fresh examples. . . . If in addition to the taste for physical well-being a social condition be added in which neither laws nor customs retain any person in his place, there is a great additional stimulant to his restlessness of temper. Men will then be seen continually to change their track for fear of missing the shortest cut to happiness. . . . When all the privileges of birth and fortune are abolished, when all professions are accessible to all, and a man's own energies may place him at the top of any one of them, an easy and unbounded career seems open to his ambition and he will readily persuade himself that he is born to no common destinies. But this is an erroneous notion, which is corrected by daily experience. . . . They have swept away the privileges of some of their fellow creatures which stood in their way, but they have opened the door to universal competition. . . . This constant strife between the inclination springing from the equality of condition and the means it supplies to satisfy them harasses and wearies the mind." (Alexis de Tocqueville, *Democracy in America*, Volume II, Book II, Chapter 12, 136-38) More than thirty years ago, I found that of sixty-nine settlers in a typical agri-

cultural district of the United States only twenty-three had any farming experience; the others included two circus musicians, three blacksmiths, two divers, two carpenters, two butchers, three cowherds, one ship's machinist, three publicans, and three old maids. (Wilhelm Röpke, "Das Agrarproblem der Vereinigten Staaten," *Archiv für Sozialwissenschaft und Sozialpolitik*, 58, p. 492)

5. On Frédéric Le Play, see my *Civitas Humana*, 111.

6. *Zahme Xenien*, V.

The inscription on Loyola's tomb in the Church of the Gesù in Rome is by an unknown author and reads: *Non coerceri maximo, contineri tamen a minimo, divinum est.* (I must thank Dr. Franz Seiler and Dr. Erik von Kuehnelt-Leddihn for this information.) Hölderlin used it, with a slight alteration, as a motto for his *Hyperion*.

7. John Stuart Mill, *On Liberty*, Chapter V; similarly, Gaëtano Mosca, *The Ruling Class* (New York, 1939), 143-44. The reference to Montesquieu and Mosca suggests that it would be rewarding to write a history of the concepts of centrism and decentrism, but to my knowledge this has never been done. I myself attempted an outline in my essay "Zentralisierung und Dezentralisierung als Leitlinien der Wirtschaftspolitik," in Ernest Lagler and Johannes Messner (eds.), *Wirtschaftliche Entwicklung und soziale Ordnung* (Vienna, 1952), 20ff.

8. It is difficult to separate the desirable from the undesirable. I tried to do so in 1950 in a report to the Adenauer government on German economic policy (*Ist die deutsche Wirtschaftspolitik richtig?*). I think that what I said there is still valid. On the one hand, every sympathy and encouragement are due the workers' and employees' wish to be taken into the management's confidence and to know about the company's affairs, which gives them a corresponding share of responsibility; the same can be said of their desire for protection against arbitrary treatment, as well as their wish to identify themselves with the company, any conflicts of interest about wage policy notwithstanding. On the other hand, it is necessary to

reject firmly any attempt to do away with subordination in decisions involving the success of the enterprise or to put part of the responsibility upon people who are not qualified for it by virtue of any expert knowledge, training, or talent and who assume no corresponding risks. Such claims must be resisted all the more forcefully because they often merely conceal an attempt by the trade-unions to extend their power to the company's management. Most of all is resistance indicated when the co-management system is used as the thin end of a wedge, with the intention to upset an economic order which, being a market economy, makes the market the source of the commands which the management's decisions try to interpret correctly. By far the best and most thorough exposition of this subject is Franz Böhm, "Das wirtschaftliche Mitbestimmungsrecht der Arbeiter im Betrieb," *ORDO, Jahrbuch für die Ordnung von Wirtschaft und Gesellschaft* (1951).

9. To give some idea of the consequences of the closed shop in England, we cite the case of Mr. Bonsor, which recently did at least arouse some public interest. The unfortunate man was a musician who had fallen behind in his union dues when out of work, but who was not allowed to accept a job until he had paid up his arrears. He eventually died as a casual worker. (*Time and Tide* [July 20, 1957])

10. There is a lot more to be said about economic concentration, especially with respect to the influence of taxation and company law, than I said in my earlier works *The Social Crisis of Our Time, Civitas Humana,* and *Mass und Mitte.* See also Joachim Kahl, *Macht und Markt* (Berlin, 1956). The best exposition of the influence of taxation known to me is to be found in the April, 1957 issue of *Wirtschaftsberichte der Berliner Bank,* which was devoted to "the disease of the German capital market." It is rightly pointed out that the "birth rate" of industrial firms, that is, the number of new firms, has sunk alarmingly low, which suggests that there is something fundamentally wrong with the capital market and the tax system. This adds rigidity to an already concentrated economic structure.

It may well be that nobody really wanted all this, and thoroughgoing reforms may therefore have good prospects. About the part played by advertising in fostering concentration, see my *Mass und Mitte*, 213ff. Meanwhile, the danger has been enhanced by television, which should have been a touchstone by which to prove whether man dominates technology or technology man; but even Switzerland failed this test.

11. G. Haberler, "Die wirtschaftliche Integration Europas," in *Wirtschaftsfragen der freien Welt*, 521-30; Röpke, *Internationale Ordnung—heute*, 308-17; idem., *L'économie mondiale aux XIXe et XXe siècles*, 135-62.

12. We may go so far as to suggest an inner kinship between Keynes and Picasso. Even if we did not know that they belong to the same era, we could guess it by the dehumanization which is characteristic of both. They also resemble each other in their alternation between the classical and the ultra-modern. Keynes greatly admired Picasso (R. Harrod, *The Life of J. M. Keynes* [London, 1951], 318), and Picasso himself, of course, is a Communist.

13. This idea is brilliantly developed in Daniel Villey, "Examen de conscience de l'économie politique," *Revue d'Economie Politique* (1951), 845-80.

14. The problem of the use of mathematics in economics has received scant attention. We may cite a discussion in the November, 1954 issue of *The Review of Economics and Statistics;* Ludwig von Mises, *Human Action* (New Haven, 1949), 347-54; and G. I. Stigler, *Five Lectures on Economic Problems* (London, 1950).

15. The discouraging experiences of Great Britain are described in detail and very frankly by Ely Devons, "Statistics as a Basis for Policy," *Lloyds Bank Review* (July, 1954). Even so conciliatory a man as D. H. Robertson says about the British planners that "the extreme inaccuracy of their forecasts ... would have had even more unfortunate consequences if the errors had not on several occasions providentially cancelled one another out." (Erik Lundberg [ed.], *The Business Cycle in the Postwar World* [London, 1955], 10) See

also Ludwig von Mises, *Theory and History* (New Haven, 1957).

16. Much confusion has been created by certain modern theories of "perfect" competition. Not only do these theories define competition in so perfectionist a manner that the necessary conditions can, *a priori*, hardly be expected to obtain in the economy, and not only has this theoretical toy nourished a pessimism which, as it were, suspects monopolistic radioactivity everywhere in the market economy, but this model of "perfect" competition also simply eliminates the dynamic nature of competition, which is precisely the basis of the arguments in favor of competition and the competitive market economy. The abstract mathematical model's concept of competition must be replaced with the concept of "active" or "workable" competition, as J. M. Clark calls it, which stresses the competitors' incessant struggle for the consumer's favor. Cf. J. M. Clark, "Toward a Concept of Workable Competition," *American Economic Review* (June, 1940); F. A. Hayek, *Individualism and Economic Order* (Chicago, 1948); Wilhelm Röpke, "Wettbewerb: Konkurrenzsystem," in *Handwörterbuch der Sozialwissenschaften*, which is to be published in the near future.

17. The great English writer Norman Angell provides an excellent illustration. In his autobiography (*After All, Autobiography of Norman Angell* [London, 1951], 102) he tells the story of how there was, at a certain stage, a proposal that he take over a Paris paper for which he had previously been working, and thereby become an entrepreneur. But he suddenly developed a "most appalling funk" because of the responsibility for all those whose livelihood would then depend upon him. When things went wrong for the paper, the kicks would be for him; when they went right, he would be regarded as a "capitalist exploiter."

18. The statement about Pitt is quoted from Bagehot, *op. cit.*, 131, and that about Alexander von Humboldt from *Briefwechsel und Gespräche Alexander von Humboldts mit einem jungen Freunde* (Berlin, 1861), 137.

INDEX

303

Market economy *cont.*
boredom of mass society, 87; part of a general order, 90-91, 93, 113, 124, 125; bourgeois foundation of, 98; self-interest in, 121; competition in, 127; leadership of, 130; restraints on by government, 141; in underdeveloped countries, 186-87; capital aid in, 188; wage level and productivity, 207; statistics in, 208; salary earners in, 240-41; small business in, 241-42; danger to, 261

Marshall Plan, 22-25

Martini, W., 270 n. 18

Marx, Karl, 21, 24, 50, 280 n. 18, 298 n. 3

Masani, M. R., 277 n. 10

Mass culture: defined, 57; downward curve of, 58; and reading of children, 58, 61; opposed to good taste, 58; and illiteracy, 59; opposed to European cultural tradition, 62; effect on elite culture, 65

Mass democracy: described as Jacobin, 66; distinguished from liberal democracy, 66; and freedom, 66; recognizes no authority above the people, 68; compared to an army, 70

Mass society: in modern life, 36; crucial issue of, 38; and Communism, 38; evidences of, 39; and city life, 39, 77; effect of on American education, 39; and collectivism, 41; population growth in, 42; individual in, 52-53, 71; and federalism, 67; production and consumption in, 72; opposed to bourgeois system, 99

Mass state: acute stage transitory, 53; acute stage causes hyperthymia, 53; chronic condition of, 54;

Mass state—*cont.*
intellectual effect of, 54-55; effect of on social structure, 55; results in urbanization, industralization, and proletarianization, 56; aided by communications media, 56; epidemic characteristics of, 57

Mass und Mitte, by Wilhelm Röpke, 8, 35, 225, 266 n. 7, 281 n. 26, 285-86 n. 6, 300 n. 10

Materialism, 108

Mathematics: limits of in economics, 248

Medical profession, 127

Melville, Herman, 58, 267 n. 8

Messner, Johannes, 299 n. 7

Meyer, F. W., 278 n. 16

Meyer, Peter, 274 n. 31, 275 n. 33

Mill, John Stuart, 50-51, 58, 63, 70, 235, 267 n. 8, 299 n. 7

Mises, Ludwig von, 301 n. 14

Mobilism, 73

Molière (*pseudonym* of Jean Baptiste Poquelin), 120

Mommsen, Theodor, 277 n. 11

Monopoly, 128

Montaigne, Michel Eyquem de, 259

Montesquieu, Baron de La Brède et de, 18, 143, 226, 235, 244, 299 n. 7

Montgomery, Field Marshal, Bernard L., 163

Morgan, Charles, 80, 181, 271 n. 22, 291 n. 18

Mosca, Gaëtano, 299 n. 7

Mother's Day: commercialism of, 129

Mowrer, Edgar Ansel, 52

Müller-Armack, A., 278 n. 16

Munby, D. L., 276 n. 4

Muschg, Walter, 269 n. 16

Mussolini, Benito, 183

Myrdal, G., 291 n. 21

Napoleon I: made Bank of France inviolable, 218